δN Formalism in Cosmological Perturbation Theory

Other Related Titles from World Scientific

Astrophysics and the Evolution of the Universe
2nd Edition
by Leonard S Kisslinger
ISBN: 978-981-3147-09-6
ISBN: 978-981-3147-10-2 (pbk)

Introduction to the Theory of the Early Universe: Hot Big Bang Theory
2nd Edition
by Valery A Rubakov and Dmitry S Gorbunov
ISBN: 978-981-3209-87-9
ISBN: 978-981-3209-88-6 (pbk)

Introduction to the Theory of the Early Universe: Cosmological
Perturbations and Inflationary Theory
2nd Edition
by Valery A Rubakov and Dmitry S Gorbunov
ISBN: 978-981-3275-62-1
ISBN: 978-981-3276-69-7 (pbk)

δN Formalism in Cosmological Perturbation Theory

Theory

Ali Akbar Abolhasani
Sharif University of Technology, Iran

Hassan Firouzjahi
Institute for Research in Fundamental Sciences (IPM), Iran

Atsushi Naruko
Tohoku University, Japan

Misao Sasaki
Kavli IPMU (WPI), University of Tokyo, Japan

World Scientific

NEW JERSEY · LONDON · SINGAPORE · BEIJING · SHANGHAI · HONG KONG · TAIPEI · CHENNAI · TOKYO

Published by

World Scientific Publishing Co. Pte. Ltd.

5 Toh Tuck Link, Singapore 596224

USA office: 27 Warren Street, Suite 401-402, Hackensack, NJ 07601

UK office: 57 Shelton Street, Covent Garden, London WC2H 9HE

Library of Congress Cataloging-in-Publication Data

Names: Abolhasani, Ali Akbar, author.

Title: Delta N formalism in cosmological perturbation theory / by Ali Akbar Abolhasani
    (Sharif University of Technology, Iran) [and three others].

Description: New Jersey : World Scientific, 2018. | Includes bibliographical references and index.

Identifiers: LCCN 2018017445| ISBN 9789813238756 (hardcover : alk. paper) |
    ISBN 9813238755 (hardcover : alk. paper)

Subjects: LCSH: Perturbation (Astronomy) | Celestial mechanics. | Cosmology.

Classification: LCC QB361 .D45 2018 | DDC 521/.4--dc23

LC record available at https://lccn.loc.gov/2018017445

British Library Cataloguing-in-Publication Data

A catalogue record for this book is available from the British Library.

For any available supplementary material, please visit
https://www.worldscientific.com/worldscibooks/10.1142/10953#t=suppl

Desk Editor: Ng Kah Fee

Typeset by Diacritech Technologies Pvt. Ltd.
Chennai - 600106, India

Printed in Singapore

Contents

Acknowledgements

We would like to thank all of our collaborators in various works involving the subject of this review, δN formalism. H. F. would like to thank YITP for hospitalities where this work was under its long preparation. This work is supported in part by MEXT grant Nos. 15H05888 and 15K21733. A.N. is supported in part by a JST grant "Establishing a Consortium for the Development of Human Resources in Science and Technology". The work of A.N. is also partly supported by the JSPS Research Fellowship for Young Scientists No.263409 and JSPS Grant-in-Aid for Scientific Research No.16H01092. We thank R. Abolhasani for the assistance in designing the monograph's cover.

CHAPTER 1

Introduction

Thanks to tremendous progresses in observational cosmology during about past two decades, see e.g. Refs. [1–7], the theory of primordial inflation has become a widely accepted, almost standard model of the early universe. Inflation [8–10] was originally proposed to solve a couple of fundamental problems associated with the standard Big Bang cosmology, namely the flatness problem that the universe is spatially so flat while the Einstein equation alone naturally suggests a highly curved universe, and the horizon problem that the universe appears so homogeneous and isotropic on large scales while there were potentially an extremely large number of causally disconnected regions in the early universe over such large scales. Through various studies, it has become clear that qualitatively inflation can be regarded as a solution to these problems, but not in the sense of quantitative, testable predictions.

It has turned out that among predictions of inflation, observationally the most important one is the generation of cosmological perturbations from quantum vacuum fluctuations that seed all large scale structures and inhomogeneities in the universe. During inflation, the quantum vacuum fluctuations of a scalar field which drives inflation, the inflaton, or those of any light scalar fields, on very small scales are stretched to very large scales due to the accelerated expansion of the universe and behave as classical fluctuations when the wavelength exceeds the Hubble horizon size. One of them or a combination of them gives rise to the curvature perturbation, namely, perturbations in the spatial curvature of the universe around the homogeneous and isotropic background Friedmann–Lemaître–Robertson–Walker (FLRW) universe. The curvature perturbation leaves a distinct imprint on the temperature anisotropy of the cosmic microwave background (CMB), and eventually develops into seed density perturbations for the large scale structure (LSS) of the universe such as galaxies and clusters of galaxies.

In this situation, it is very important to quantify theoretical predictions of various inflationary models in detail, and test them against the observed CMB anisotropy data and against various observations of the large scale structure of the universe. For this purpose, the relativistic cosmological perturbation theory plays the central role. In particular, in the leading order, linear perturbation theory has turned out to be extremely powerful. Nevertheless, in the last decade much attention has been paid to the non-linear nature encoded in the primordial perturbations since it is now within the scope of observations. In particular, such non-linearities generate deviations from the Gaussian statistics, that is, non-Gaussianities even if primordial fluctuations originate from vacuum fluctuations with the Gaussian statistics [11]. As a result, any detection of primordial non-Gaussianities through the statistics of CMB anisotropy or distributions of LSS may give a valuable clue to decipher the detailed mechanism of primordial inflation, e.g. non-linear interaction among fields.

The simplest class of inflationary models is the single-field slow-roll model, in which a single scalar field, the inflaton field ϕ, is responsible for both driving inflation and generating the curvature perturbations. One of the most remarkable fact of primordial fluctuations from single-field inflation is that one of the modes of the spatial curvature perturbation on the velocity-orthogonal hypersurface (misleadingly but commonly dubbed as "comoving" hypersurface), namely the hypersurface on which the scalar field is homogeneous, is conserved after the mode exits the Hubble horizon [12, 13].

To be specific, the spatial curvature on the comoving hypersurface at linear order may be expressed as

$$R^{(3)}\big|_{\text{comoving}} = -\frac{4}{a^2}\delta^{ij}\partial_i\partial_j\mathcal{R}_c, \qquad (1.1)$$

where $\mathcal{R}_c$ is the comoving curvature perturbation and a is the cosmological scale factor. It is known that $\mathcal{R}_c$ satisfies the following second order differential equation,

$$\mathcal{R}_c'' + 2\frac{z'}{z}\mathcal{R}_c' - \delta^{ij}\partial_i\partial_j\mathcal{R}_c = 0, \qquad (1.2)$$

in which the quantity z is related to the inflaton field ϕ via $z \equiv a(\phi'/\mathcal{H}) = a(\dot{\phi}/H)$. Here $'$ denotes a derivative with respect to the conformal time, $d\eta = a^{-1}dt$ while t represents the cosmic time and $\mathcal{H} = a'/a$ is the conformal Hubble parameter related to the standard Hubble parameter H via $\mathcal{H} = aH$. On sufficiently large scales, one can safely neglect the last term in Eq. (1.2) which involves spatial gradients and hence $\mathcal{R}_c' \sim 0$ becomes one of the solutions of this equation. This conserved mode is called the adiabatic growing mode (for a reason we will explain later).

Furthermore, under the slow-roll approximation, z essentially behaves as the scale factor a and then the other solution gives $\mathcal{R}_c' \propto z^{-2} \sim a^{-2}$ which soon becomes negligible once the mode is outside the Hubble horizon scale during inflation. So the comoving curvature perturbation is indeed conserved on superhorizon scales.[1] This fact leads to another important consequence. In principle, there are two dynamical degrees of freedom in single-field inflation since the scalar field satisfies a second order differential equation, and the two physical degrees of freedom are imprinted in the field value ϕ and its time derivative $\dot{\phi}$ at a given initial time. However, on large scales, one of these modes decays out and only one dynamical degree of freedom

---

[1]We mention here that it is customary to call the scales which exceed the Hubble radius as superhorizon scales. We adopt this terminology throughout this monograph, though more rigorously speaking perhaps they should be called super-Hubble or super-Hubble horizon scales.

survives. Reflecting this fact, one finds the coincidence of several time slicings on large scales.

One can always consider a slicing where the energy density of the universe is spatially homogeneous, which is called the uniform energy density slicing. Since the energy density ρ of the scalar field is given by the sum of the potential and kinetic energies, its value can be determined by specifying both ϕ and $\dot{\phi}$. However, it turns out that the energy density may be regarded as a function of ϕ alone due to the reduction of the two degrees of freedom to a single degree of freedom. Hence on the comoving slice on which ϕ is uniform, one notices that the energy density ρ is also uniform, which implies the coincidence of the two slicings. Moreover, the Friedmann equation, or more precisely the Hamiltonian constraint equation which includes perturbations, equates the energy density with the Hubble expansion rate H on large scales. Then one also finds the coincidence of the above two slicings with the uniform Hubble (expansion rate) slicing.

To summarize, we have

$$\text{comoving} \quad \sim \quad \text{uniform energy} \quad \sim \quad \text{uniform Hubble} \qquad (1.3)$$

on large (superhorizon) scales. Based on this fact, one notices that the conservation of the curvature perturbation is not a special property of the comoving slicing. Even for other slicings, such as the uniform energy density or uniform Hubble slicings, the curvature perturbation may be also conserved. This is a rather generic physical consequence of single-field slow-roll inflation.

On the other hand, the conservation of the curvature perturbation no longer holds in multi-field inflation models [14, 15]. Even under the slow-roll approximation under which all the decaying modes may be neglected, there still remain multiple dynamical degrees of freedom. These dynamical degrees of freedom of multi-scalar fields affect the evolution of the curvature perturbation even on superhorizon scales. Those effects appear on the right-hand side of Eq. (1.2) as a source term. This may be seen by writing down the energy conservation law of a perfect fluid on uniform density slicing. On superhorizon scales, it reads

$$\mathcal{R}'_e \simeq -\frac{\mathcal{H}}{\rho+P}\delta P_{\text{nad}} \qquad (1.4)$$

where $\mathcal{R}_e$ is the curvature perturbation on uniform density slices, $\delta P_{\text{nad}} \equiv \delta P - \frac{\dot{P}}{\dot{\rho}}\delta\rho$ is the non-adiabatic pressure perturbation, and we have neglected spatial gradients by adopting the large-scale approximation. It is clear that the adiabatic pressure perturbation vanishes on the uniform energy density slicing because the adiabatic pressure means $P = P(\rho)$. So any non-vanishing pressure perturbation on the uniform density slicing is non-adiabatic, and it seeds the curvature perturbation.

In multi-field inflation, δP_{nad} may be a complicated combination of the multi-field components, whose evolution equations may be coupled to each other. Thus, although one may be able to find the time evolution of δP_{nad} and then that of the curvature perturbation by carefully solving the perturbation equations, it may not be an easy task.

The δN formalism is invented to save this situation [16–18]. It is based on the fact that on superhorizon scales each Hubble horizon size region evolves mutually independently because they are not causally connected to each other [15, 19–21]. It is essentially equivalent to the leading order approximation in spatial gradient expansion, and it is often called the separate universe approach. It turns out that the comoving curvature perturbation can be computed by solving the equations for the homogeneous background with slightly different initial conditions, without resorting to the complicated perturbation equations.

In this monograph, we review the δN formalism in depth. It is organized as follows. In Chapter 2 we formulate the δN formalism at linear and non-linear order. First the δN formalism is introduced at linear order in cosmological perturbation theory. The derivation and analysis of the linear δN formalism are quite useful because one will be able to clearly understand the geometrical meaning of the δN formalism, and explicitly observe the correspondence between a set of background homogeneous equations/solutions and those for perturbations. Then we derive the non-linear δN formalism by extending the definition of the comoving curvature perturbation to full non-linear order, and discuss the conservation of the thus-defined non-linear curvature perturbation.

In Chapter 3 we present a warmup application of δN formalism in simple models of inflation, such as the single-field chaotic inflation. In addition, as another interesting application of δN formalism, in this chapter we review the curvaton scenario. These examples are good starting points where δN analysis are simple before we look for more involved models of inflation with non-trivial δN analysis in the following chapters.

In Chapter 4, as an application of the δN formalism, we consider the model of multi-brid inflation, which is a multi-field version of hybrid inflation. First we introduce a class of analytically soluble models. Then we focus on multi-brid inflation because it allows us to compute the curvature perturbation at linear and non-linear orders analytically, and hence it is easy to see how the δN formalism works.

In Chapter 5 we apply the δN formalism to another intriguing example: non-attractor models of inflation. As mentioned before, the slow-roll approximation generally kills half of the dynamical degrees of freedom because they correspond to decaying modes. However, there are cases where these would-be decaying modes are not actually decaying but growing. In such cases one cannot use the slow-roll approximation, and in particular one has to take both degrees of freedom of ϕ and $\dot{\phi}$ (i.e. two independent degrees of freedom) in the analysis even in single-field inflation. They are called non-attractor inflation because the trajectories in the phase space have not converged to a single, attractor trajectory along which, say, $\dot{\phi}$ is no longer independent of ϕ.

Finally, in Chapter 6 we present the application of δN formalism in models of inflation with local features. Here local features mean the special features localized in modes with specific length scales. This is a subset of the multi-field scenarios in which the heavy field becomes tachyonic during inflation inducing local effects in the power spectrum and bispectrum of curvature perturbations. This provides yet another non-trivial example of applicability of δN formalism which may be difficult to incorporate in the alternative perturbation theory methods.

Some comments are in order before we close this chapter. First, our convention for the metric signature is $(-,+,+,+)$. Second, we introduce different sets of slow-roll parameters in different chapters. One set of slow-roll parameters is defined via the variation of the Hubble parameter defined as $\epsilon \equiv -\frac{\dot{H}}{H^2}$, $\eta \equiv \frac{\dot{\epsilon}}{H\epsilon}$. The second definition is given via the derivatives of the potential as $\epsilon_V \equiv \frac{M_P^2}{2}(\frac{V_\phi}{V})^2$, $\eta_V \equiv M_P^2 \frac{V_{\phi\phi}}{V}$. In each chapter only one definition of slow-roll parameters is used so there would be no confusion. Third, as mentioned above, in this monograph we mainly focus on δN formalism and its applications in various inflationary scenarios. Therefore, only the references directly related to δN formalism and its specific follow up applications are mentioned. There are huge literature on various aspects of inflationary model buildings and cosmological perturbation theory which unfortunately are not mentioned in our list of references. While this monograph deals mainly with δN formalism, there are good textbooks and review papers on inflation and cosmological perturbation theory based on standard approach [22–29]. We refer the reader to these textbooks and reviews and the references therein for alternative reviews on cosmological perturbation theory.

CHAPTER **2**

Basic formulation of δN formalism

2.1 Preliminary

2.1.1 *Linear perturbation theory*

Let us first briefly summarize some of the important equations and results from linear cosmological perturbation theory. We consider perturbations around a spatially flat Friedmann–Lemaître–Robertson–Walker (FLRW) background. Since the spatial section of an FLRW metric is homogeneous and isotropic, one can decompose the perturbation into scalar, vector and tensor quantities with respect to the symmetry of the 3-space. A perturbation that can be described by a scalar quantity is called the scalar-type, that by a vector the vector-type, and that by a tensor the tensor-type perturbation. In general relativity, the tensor-type perturbations contain the true geometrical degrees of freedom of the metric and they correspond to gravitational wave modes. On the other hand, there are no intrinsic physical degree of freedom in either scalar-type or vector-type perturbations. However, once we include matter, the scalar-type or vector-type perturbations may become physical, depending on the nature of the matter.

In this monograph, we focus on scalar-type perturbations, to which the δN formalism may be applied. Then we may define the metric perturbation variables based on Ref. [22] as

$$ds^2 = g_{\mu\nu}dx^\mu dx^\nu = \left(\bar{g}_{\mu\nu} + \delta g_{\mu\nu}\right)dx^\mu dx^\nu$$
$$= a^2\left(\eta\right)\left[-\left(1+2A\right)d\eta^2 - 2(-\triangle)^{-1/2}B_{,i}dx^i d\eta\right.$$
$$\left. + \left\{\left(1+2\mathcal{R}\right)\delta_{ij} - 2\left(-\triangle\right)^{-1}C_{,ij}\right\}dx^i dx^j\right],$$

(2.1)

where $\delta g_{\mu\nu}$ denotes the metric perturbation and $\bar{g}_{\mu\nu}$ the background metric. For convenience we have introduced the conformal time defined by $d\eta = dt/a\left(t\right)$ in terms of the cosmic time, t. We have introduced the symbol $\triangle$ for the 3-dimensional Laplacian, $\triangle \equiv \delta^{ij}\partial_i\partial_j$, and $(-\triangle)^{-1/2}$ and $\triangle^{-1}$ are non-local operators which have their obvious meanings, respectively, defined through the Fourier transform of the variables. These non-local operators are inserted to make the metric perturbation variables B and C dimensionless. Unless otherwise noted, we choose the spacetime coordinates as $x^\mu = \left(\eta, x^i\right)$ in what follows, where μ runs from 0 to 3 while i from 1 to 3.

The geometrical meaning of each variable introduced above may be made clearer by utilizing the notion of the $(3+1)$-decomposition of the metric. In the $(3+1)$-decomposition, A is the lapse function perturbation which describes how the time slices are embedded in the spacetime and B is the shift vector perturbation which describes how the spatial coordinates are threaded through the spacetime. The variable $\mathcal{R}$ can be related with the intrinsic spatial curvature since the spatial curvature is givenby

$$\delta R_{ij}^{(3)} = -\left(\partial_i\partial_j + \delta_{ij}\triangle\right)\mathcal{R}, \qquad \delta R^{(3)} = -\frac{4}{a^2}\triangle\mathcal{R}.$$

(2.2)

Therefore we simply call $\mathcal{R}$ as the curvature perturbation. The remaining variable C represents anisotropy in the spatial metric, and its time derivative accompanied with the shift vector represents the shear of the unit vector field normal to the time-constant hypersurface [22],

$$(-\triangle)^{1/2}\sigma_g \equiv C' - (-\triangle)^{1/2}B,$$

(2.3)

where $'$ denotes a derivative with respect to the conformal time.

As for the matter sector, the energy momentum tensor must be in the form of a perfect fluid on the FLRW background in order for it to be compatible with the assumption of spatial homogeneity and isotropy,

$$T^{\mu\nu} = \rho u^\mu u^\nu + P\left(u^\mu u^\nu + g^{\mu\nu}\right), \tag{2.4}$$

where u^μ is the unit time-like vector along which the energy flows (the four-velocity in the case of a fluid), i.e. it is subject to the normalization condition $g_{\mu\nu}u^\mu u^\nu = -1$. Besides, ρ and P are the energy density and pressure, respectively, in the rest frame of u^μ. Then the perturbation of the energy momentum tensor, δT^μ_ν can be expressed as

$$\delta T^0_{\ 0} = -\delta\rho, \tag{2.5a}$$

$$\delta T^0_{\ j} = -\left(\rho + P\right)\left(-\triangle\right)^{-1/2}\partial_j\left(v + B\right), \tag{2.5b}$$

$$\delta T^i_{\ j} = \delta P \delta^i_j - \left[\left(-\triangle\right)^{-1}\partial^i\partial_j + \frac{1}{3}\delta^i_j\right]\Pi, \tag{2.5c}$$

where $\delta\rho$ and δP are the energy density and pressure ($=$ isotropic stress) perturbations respectively; v represents the velocity perturbation,

$$u^i/u^0 = -\left(-\triangle\right)^{-1/2}\partial^i v, \tag{2.5d}$$

where $\partial^i = \delta^{ij}\partial_j$; and Π denotes the anisotropic stress perturbation.

So far we have considered the most general form of the scalar-type perturbations. We now confine our attention to general relativity (or more generally any gravity theory endowed with general covariance). Then an infinitesimal coordinate transformation,

$$x^\mu \to \overline{x}^\mu = x^\mu + \xi^\mu, \tag{2.6}$$

induces a gauge transformation in linear perturbation theory,

$$\delta g_{\mu\nu} \to \delta\overline{g}_{\mu\nu} = \delta g_{\mu\nu} - \xi_{\mu;\nu} - \xi_{\nu;\mu}, \tag{2.7}$$

$$\delta T^\mu_{\ \nu} \to \delta\overline{T}^\mu_{\ \nu} = \delta T^\mu_{\ \nu} - T^\mu_{\ \nu;\alpha}\xi^\alpha - T^\mu_{\ \alpha}\xi^\alpha_{;\nu} + \xi^\mu_{;\alpha}T^\alpha_{\ \nu}. \tag{2.8}$$

For the scalar-type perturbations, there are two gauge degrees of freedom which may be expressed as

$$\xi^\mu = \left(T, -(-\triangle)^{-1/2}\partial^j L\right), \tag{2.9}$$

where T and L are arbitrary functions.

Under this gauge transformation, the metric perturbation variables transform as

$$A \quad \to \quad \overline{A} = A - T' - \mathcal{H}T, \tag{2.10a}$$

$$B \quad \to \quad \overline{B} = B - L' - (-\triangle)^{1/2}T, \tag{2.10b}$$

$$C \quad \to \quad \overline{C} = C - (-\triangle)^{1/2}L, \tag{2.10c}$$

$$\mathcal{R} \quad \to \quad \overline{\mathcal{R}} = \mathcal{R} - \mathcal{H}T, \tag{2.10d}$$

where $\mathcal{H} = a'/a\,(= aH)$ is the conformal Hubble parameter, while $H = \dot{a}/a$ is the conventional (proper time) Hubble parameter. One also finds the gauge transformation rule for the shear as

$$\sigma_g \to \overline{\sigma}_g = \sigma_g + (-\triangle)^{1/2}T. \tag{2.10e}$$

As is clear from the above, the function T reflects the degree of freedom in the choice of time slicing, while the function L represents the choice of the spatial coordinates.

The matter variables transform as

$$\delta\rho \quad \to \quad \delta\overline{\rho} = \delta\rho + 3\mathcal{H}(\rho + P)T, \tag{2.11a}$$

$$\delta P \quad \to \quad \delta\overline{P} = \delta P + 3c_w^2\mathcal{H}(\rho + P)T, \tag{2.11b}$$

$$v \quad \to \quad \overline{v} = v + L', \tag{2.11c}$$

$$v + B \quad \to \quad \overline{v} + \overline{B} = v + B - (-\triangle)^{1/2}T. \tag{2.11d}$$

Here c_w^2 is defined by

$$c_w^2 \equiv \frac{P'}{\rho'} = \frac{\dot{P}}{\dot{\rho}}, \tag{2.11e}$$

and it corresponds to the adiabatic sound velocity. On the other hand, it should be noted that in the case of a scalar field the speed of sound c_s^2 does not coincide with c_w^2. For example $c_s^2 = 1$ in the case of a canonical scalar field while c_w^2 is not unity. It also should be noted that the anisotropic stress perturbation Π is gauge-invariant because its background value is zero.

In what follows, to be specific, we consider the Einstein equation,

$$G_{\mu\nu} = 8\pi G T_{\mu\nu}, \qquad (2.12)$$

where G is the gravitational constant. The dynamics of the homogeneous and isotropic background universe is governed by the Friedmann equation and the energy conservation equation,

$$3H^2 = \frac{1}{M_P^2}\,\rho\,, \quad \dot{\rho} + 3H\left(\rho + P\right) = 0\,, \qquad (2.13)$$

where M_P is the Planck mass related to G as $8\pi G = M_P^{-2}$ in natural units $\hbar = c = 1$.

Then using the perturbation variables introduced above in Eqs. (2.1) and (2.5), one can explicitly write down the perturbed Einstein equations. To solve them, one needs to fix gauges in general. Equivalently it is also possible to take the gauge-invariant approach in which one writes down the perturbed Einstein equations in terms of gauge-invariant variables alone, i.e., combinations of the variables that are invariant under the gauge transformations presented above. However, since a gauge-invariant variable may be always regarded as a quantity defined in a certain gauge in which the gauge fixing is completed, we will not take that approach in this monograph.

The most important task in fixing the gauges is to specify the way to choose time slicing since the time-dependence of the background is non-trivial while it bears a maximally symmetric 3-surface. Among various possible choices of the time slicing, one that happens to play a very important role in cosmological perturbation theory is called the comoving slicing. It is defined by imposing $\delta T_i^0 = 0$, i.e.,

$$v + B = 0\,. \qquad (2.14)$$

As is clear from Eq. (2.11d), this fixes the time slicing completely with no further degree of freedom in the gauge variable T. The name "comoving" slicing comes from the following consideration. If we recall the definition of v in Eq. (2.5d), we see that the covariant components of the 4-velocity are given by

$$u_i = g_{ij}u^j + g_{i0}u^0 = -a(-\triangle)^{-1/2}\partial_i(v+B). \qquad (2.15)$$

Thus $v + B = 0$ requires $u_i = 0$, which implies that the vector u^μ is orthogonal to the $t = \text{const.}$ hypersurface. Hence the rest frame of u^μ coincides with the natural rest frame of the hypersurface defined by its orthogonal direction. In this sense, the hypersurface may be regarded as being "comoving" with the matter, though it may not be an appropriate usage in the sense of conventional fluid mechanics.

In passing, we also note that the terminology "comoving slicing" should not be confused with the "comoving gauge". The word "comoving" is usually related with the choice of the spatial coordinates. The coordinates x^i are called comoving if the matter flows along fixed x^i. Thus the comoving gauge would mean $v = 0$, not $v + B = 0$. Of course one may choose a gauge in which both v and B are set to zero. In that case one may call it the comoving slicing comoving gauge in the most rigorous sense.

The important property of the comoving slicing is that the curvature perturbation on that slice, which we call the comoving curvature perturbation denoting it by $\mathcal{R}_c$, satisfies a Klein–Gordon type second-order equation without mass term. Neglecting the anisotropic stress perturbation, which is a good approximation for most models of inflation, it takes the following form

$$
\mathcal{R}_c'' \frac{d}{d\eta}\left(\ln\left[a^2\frac{\rho+P}{c_s^2\rho}\right]\right)\mathcal{R}_c' - c_s^2\triangle\mathcal{R}_c \\
= -\frac{\mathcal{H}}{\rho+P}\left(P_{\text{nad}}' + \frac{d}{d\eta}\left[\ln\left(\frac{a^3}{c_s^2\sqrt{\rho}}\right)\right]P_{\text{nad}}\right). \qquad (2.16)
$$

Here c_s is the adiabatic sound speed, which coincides with c_w introduced in Eq. (2.11e) in the case of a fluid. And P_{nad} is the non-adiabatic pressure perturbation,

$$P_{\text{nad}} \equiv \left(\delta P - c_s^2\delta\rho\right)_c, \qquad (2.17)$$

where the subscript c denotes that it be must evaluated on the comoving hypersurface. Thus in the absence of the non-adiabatic perturbation, there exists a solution on sufficiently large scales which is constant in time, i.e., conserved. In the actual universe, it is widely believed (and is perfectly consistent with observations) that the non-adiabatic perturbation becomes negligible by some epoch after inflation but well before the universe becomes dominated by dark matter. At such a stage, the universe is presumably dominated by radiation ($c_s^2 \approx 1/3$), and Eq. (2.16) on superhorizon scales reduces to

$$\mathcal{R}_c'' + 2\mathcal{H}\mathcal{R}_c' = 0. \tag{2.18}$$

Thus the constant solution dominates and $\mathcal{R}_c$ becomes conserved irrespective of its initial condition.

We say the universe has reached the adiabatic limit when the non-adiabatic perturbation has become negligible. In this limit, all the information about the nature of fluctuations produced during inflation is encoded in this conserved comoving curvature perturbation. Below we shall see that the conservation holds even at full non-linear order for an appropriately extended, non-linear version of the comoving curvature perturbation. One of the main objectives of studying cosmological perturbation theory is to compute the final, conserved value of the comoving curvature perturbation. The δN formalism has been developed for this purpose.

2.1.2 *Single-field slow-roll inflation*

The simplest inflationary scenario is the single-field slow-roll inflation in which a single scalar field, the inflaton, slowly rolls over a nearly flat potential. Thanks to the slow-roll condition, the background universe is nearly de Sitter, which leads to a set of generic predictions that the curvature perturbations are adiabatic, nearly scale-invariant and nearly Gaussian. As discussed in Chapter 1, these predictions are perfectly consistent with cosmological observations. This provides a strong motivation to study and understand the single-field slow-roll inflation.

Here we consider the δN formalism in single-field slow-roll inflation and compute the primordial curvature perturbation. To simplify the situation, we assume that the scalar field has the canonical kinetic term. Then the model is specified once the potential $V(\phi)$ is specified.

The equations that govern the background spacetime during inflation are the Friedmann equation and the scalar field equation,

$$3M_P^2 H^2 = \rho = \frac{1}{2}\dot{\phi}^2 + V(\phi),\tag{2.19}$$

$$\ddot{\phi} + 3H\dot{\phi} + V_{,\phi} = 0.\tag{2.20}$$

To begin with, we assume that the universe is in a stage of quasi-exponential expansion. This means $|\dot{H}|/H^2 \ll 1$. From the Einstein equations (or the time derivative of the Friedmann equation together with the scalar field equation), this means

$$\epsilon \equiv -\frac{\dot{H}}{H^2} = \frac{\frac{3}{2}\dot{\phi}^2}{\frac{1}{2}\dot{\phi}^2 + V} \ll 1.\tag{2.21}$$

The quantity ϵ is usually called the slow-roll parameter, as the smallness of it implies that of the kinetic term in comparison with the potential term, though this may not be necessarily true in a more general class of models with non-minimal kinetic terms. As mentioned above, the smallness of ϵ is the necessary and sufficient condition for the universe to be inflating.

A couple of important quantities which characterize models of inflation are

$$\eta \equiv \frac{\dot{\epsilon}}{H\epsilon},\tag{2.22}$$

and

$$\delta \equiv \frac{\ddot{\phi}}{H\dot{\phi}} = -\epsilon + \frac{\eta}{2},\tag{2.23}$$

where the second equality in Eq. (2.23) holds for models with the canonical kinetic term. As is clear from the definition, the smallness of δ is the condition for the field to be slow-rolling. Thus the slow-rolling implies $|\delta| \ll 1$ and since $\epsilon \ll 1$, it implies $|\eta| \ll 1$. The quantities ϵ and η are called the slow-roll parameters of inflation, though ϵ is not necessarily related to the slow-rolling of the inflaton field.

Now we consider the situation in which the inflaton field slowly rolls down the potential, and the universe is inflating. Hence we have

$|\delta| \ll 1$ and $\epsilon \ll 1$. Under these conditions, the Friedmann and field equations are approximated as

$$3M_P^2 H^2 = V(\phi), \tag{2.24}$$

and

$$\dot{\phi} = -\frac{V_{,\phi}}{3H}. \tag{2.25}$$

The condition for these approximations to be valid can be expressed in terms of the potential as

$$\epsilon_V \equiv \frac{M_P^2 V_{,\phi}^2}{2V^2} \ll 1, \quad \eta_V \equiv \frac{M_P^2 V_{,\phi\phi}}{V} \ll 1, \tag{2.26}$$

where

$$\epsilon_V \approx \epsilon, \quad \eta_V \approx -\frac{\eta}{2} + 2\epsilon = \epsilon - \delta. \tag{2.27}$$

The models that satisfy these conditions are called slow-roll inflation models. The most important nature of this class of models is that the value of ϕ plays the role of time: Eq. (2.24) implies that H is a function of ϕ, and hence Eq. (2.25) implies a one-to-one correspondence between ϕ and t,

$$dt = -\frac{3H}{V_{,\phi}} d\phi. \tag{2.28}$$

In fact there is a wider class of models which have the same nature. The essential point is that the value of the scalar field completely determines the value of its time derivative. A system where this is realized is said to be in the attractor stage. Thus slow-roll inflation can be alternatively called attractor inflation in a generic sense. On the other hand, even if ϕ is rolling very slowly in the conventional sense, $|\dot{\phi}| \ll H|\phi|$, the system may not be in an attractor stage. In this case, the evolution of the system is not solely determined by the value of ϕ, but the value of $\dot{\phi}$ becomes important as an independent degree of freedom. The models that contain such a stage are called non-attractor inflations. This complicated but interesting situation will be studied in Chapter 5.

Using the Friedmann equation and referring back to the definition of the Hubble expansion rate, $H = \dot{a}(t)/a(t)$, one can readily find the number of e-folds of inflation from the time t until the end of inflation at $t = t_f$, $a = a(t_f) = a_f$, as

$$N(t) \equiv \ln\left(\frac{a_f}{a(t)}\right) = \int_t^{t_f} H dt'. \qquad (2.29)$$

As mentioned above, in the single-field slow-roll models of inflation, the inflaton $\phi(t)$ plays the role of time. Therefore, we can change the clock from t to ϕ and change dt to $d\phi$ by Eq. (2.28) to obtain

$$N(\phi) = -\int_\phi^{\phi_f} \frac{3H^2}{V_{,\phi}} d\phi = \int_{\phi_f}^\phi \frac{3H^2}{V_{,\phi}} d\phi = \frac{1}{M_P^2} \int_{\phi_f}^\phi \frac{V}{V_{,\phi}} d\phi \qquad (2.30)$$

where the last equality follows form Eq. (2.24).

Then the δN formula is derived as follows. Let us first consider the variation of N with respect to the value of ϕ, that is, the variation of the initial time,

$$\delta N(\phi) = \frac{1}{M_P^2} \frac{V}{V_{,\phi}} \delta\phi. \qquad (2.31)$$

Since this is just an increment of N in response to the variation of the initial time ϕ, it has formally nothing to do with the curvature perturbation. However, in the current case of single-field slow-roll inflation, one can directly relate it to the comoving curvature perturbation because of the conservation on superhorizon scales.

To show this, we consider a transformation from the flat slicing to the comoving slicing. Let $\Delta t_{\mathrm{fl}\to c}$ be the infinitesimal time difference necessary to transform from the flat to comoving slicings. Then the gauge transformation rules tell us that

$$\begin{aligned} \delta\phi_c &= \delta\phi_{\mathrm{fl}} - \dot{\phi}\Delta t_{\mathrm{fl}\to c}, \\ \mathcal{R}_c &= \mathcal{R}_{\mathrm{fl}} - H\Delta t_{\mathrm{fl}\to c}, \end{aligned} \qquad (2.32)$$

where $\delta\phi_c = \mathcal{R}_{\mathrm{fl}} = 0$ by definition. Thus we immediately obtain

$$\mathcal{R}_c(t_i) = -H\Delta t_{\mathrm{fl}\to c}(t_i) = -\frac{H}{\dot{\phi}}\delta\phi_{\mathrm{fl}}(t_i) = \frac{1}{M_P^2} \frac{V}{V_{,\phi}} \delta\phi_{\mathrm{fl}}(t_i), \qquad (2.33)$$

where t_i is the initial time when the scale of interest has just exceeded the Hubble horizon scale. Now since $\mathcal{R}_c$ is conserved on superhorizon scales, we have $\mathcal{R}_c(t_f) = \mathcal{R}_c(t_i)$ for any t as far as the scale is superhorizon. Thus comparing Eqs. (2.31) and (2.33), we find

$$\mathcal{R}_c(t_f) = \delta N = \frac{1}{M_P^2} \frac{V}{V_{,\phi}} \delta \phi_{\text{fl}}(t_i, \mathbf{x}) . \qquad (2.34)$$

Namely, the conserved comoving curvature perturbation we want to compute is given by the variation of the number of e-folds due to the variation of the inflaton fluctuation on flat slicing evaluated at an initial time. This is the δN formula for the single-field slow-roll inflation.

Most commonly, the above δN formula is used in the Fourier space, by focusing on a single comoving wavenumber $\mathbf{k}$. In this case the most convenient choice of the initial time is when the physical wavelength has just exceeded the horizon size, or the horizon-crossing time, $t = t_k$ such that $k = H(t_k) a(t_k)$. To be rigorous, it should be chosen when the wavelength has become sufficiently larger than the horizon size, which happens after a few number of e-folds from the horizon-crossing time. However, thanks to the slow-roll assumption, the time variation of the fluctuation $\delta \phi_{\text{fl}}(t, \mathbf{k})$ is small enough so that its variation during several e-folds after horizon-crossing can be neglected. Thus approximating it by the value at horizon-crossing is valid with good accuracy. For notational simplicity, let us denote it with an asterisk: $\delta \phi_*(\mathbf{k}) \equiv \delta \phi_{\text{fl}}(t_k, \mathbf{k})$. Thus Eq. (2.34) becomes

$$\mathcal{R}_c(t_f, \mathbf{k}) = \frac{1}{M_P^2} \frac{V}{V_{,\phi}} \bigg|_{t_k} \delta \phi_*(\mathbf{k}) . \qquad (2.35)$$

Since we have assumed that the universe is already in the adiabatic stage during inflation, the comoving curvature perturbation is conserved for all time until it re-enters the horizon. Hence $\mathcal{R}_c(t_f, \mathbf{k})$ is in fact independent of t_f as long as it is on superhorizon scales. Hence we may remove the argument t_f from it and simply denote it by $\mathcal{R}_c(\mathbf{k})$, with the understanding that it is the conserved comoving curvature perturbation on superhorizon scales.

In what follows, we show that the above formula can be generalized to multi-field models as well as to the full non-linear case, even though the comoving curvature perturbation would no longer be conserved on superhorizon scales in general.

2.2   δN formalism in linear perturbation theory

2.2.1   *δN formalism in slow-roll inflation*

δN formula and curvature perturbation

Here we first introduce the notion of the number of *e*-folds in general spacetime, $\mathcal{N}$, and then define the perturbation of it, δN, with respect to an (unperturbed) FLRW background, N. Let us take the $(3+1)$-decomposition approach in which spacetime is foliated by slices of space-like hypersurfaces, $t = \text{const.}$ (or $\eta = \text{const.}$). We can then thread the slices with trajectories orthogonal to the $t = \text{const.}$ hypersurfaces (but the spatial coordinates need not be constant along these trajectories). Let us denote the unit normal to $t = \text{const.}$ by n^{μ}. We define the number of *e*-folds, $\mathcal{N}$, from $t = t_1$ to t_2 $(t_1 \leq t_2)$ by integrating the expansion along a trajectory orthogonal to the slices,

$$\mathcal{N}(t_1, t_2) = \frac{1}{3} \int_{\tau_1}^{\tau_2} K \, d\tau; \quad K \equiv \nabla_{\mu} n^{\mu}, \tag{2.36}$$

where K is the expansion and τ is the proper time along the trajectory whose precise definition will be given in (2.147).

In an FLRW universe, the natural time slices are those that respect the spatial homogeneity and isotropy. Then K is given by the Hubble expansion rate H as $K = 3H$, and the cosmic time t coincides with the proper time τ along the trajectories orthogonal to the slices. In this case the number of *e*-folds reduces to the standard one defined in cosmology,

$$N(t_1, t_2) = \int_{t_1}^{t_2} H \, dt = \ln \left[\frac{a(t_2)}{a(t_1)} \right], \tag{2.37}$$

where $a(t)$ is the cosmic scale factor.

Applying the general definition of $\mathcal{N}$ in Eq. (2.36) to the linearly perturbed metric (2.1), we find

$$K = \frac{1}{a} \left[3\mathcal{H}(1 - A) + 3\mathcal{R}' + (-\triangle)^{1/2} \sigma_g \right], \tag{2.38}$$

and

$$d\tau = a\left(1 + A\right)d\eta. \tag{2.39}$$

Then the perturbed number of e-folds is expressed as

$$\mathcal{N} = N + \int_{\eta_1}^{\eta_2} d\eta \left[\mathcal{R}' + \frac{1}{3}(-\triangle)^{1/2}\sigma_g\right]. \tag{2.40}$$

The shear σ_g decays very quickly on large scales where spatial gradients can be neglected. As may be shown by using the traceless part of the Einstein equations (2.65),

$$\sigma_g \propto a^{-2}. \tag{2.41}$$

Thus we can neglect σ_g in Eq. (2.40) to obtain

$$\mathcal{N} \simeq N + \int_{\eta_1}^{\eta_2} d\eta \, \mathcal{R}' = N + \int_{t_1}^{t_2} dt \, \dot{\mathcal{R}} \tag{2.42}$$

$$= N + \mathcal{R}\left(t_2\right) - \mathcal{R}\left(t_1\right). \tag{2.43}$$

Therefore in terms of the perturbed number of e-folds, we have

$$\delta N \equiv \mathcal{N} - N = \mathcal{R}\left(t_2\right) - \mathcal{R}\left(t_1\right). \tag{2.44}$$

This means that the perturbation in the number of e-folds is equal to the difference between the curvature perturbation on the final time slice $(t = t_2)$ and that on the initial time slice $(t = t_1)$. Note that this result holds irrespective of choice of slicing.

We also mention that Eq. (2.44) may be regarded as a purely geometrical relation, practically independent of theory of gravity behind. This is because of the fact that the only role of the Einstein equation in the above is to show the rapid decay of σ_g, but such behavior is quite general in any metric theory of gravity. In fact, unless the matter sector allows a sustainable anisotropy on infinitely large scales, which seems to be guaranteed for most models of the universe which are dominated by a scalar field or by a fluid, the rapid decay of σ_g is a common feature in any theory of gravity as long as it is healthy (i.e., it has no physical instability on superhorizon scales).

As discussed in the previous subsection, the most important quantity to compute is the final, conserved comoving curvature perturbation in the adiabatic limit. So let us identify the time t_2 with an epoch after the universe has reached the adiabatic limit and denote it by t_f. We say the universe has reached the adiabatic limit when the non-adiabatic perturbation, as well as the decaying mode of the above equation, has become negligible. Namely, we choose $\mathcal{R}(t_2) = \mathcal{R}_c(t_f)$. As for the initial time slice, we impose a different condition for the time slicing, that is, the vanishing of $\mathcal{R}$ on the time hypersurface.

We call it the flat slice because the spatial curvature, $\mathcal{R}$, vanishes on that slice as can be seen from Eq. (2.2). Namely replacing t_1 by t_i, we set $\mathcal{R}(t_i) = 0$. Then Eq. (2.44) reduces to

$$\delta N\left(t_i, t_f\right) = \mathcal{R}_c\left(t_f\right) . \tag{2.45}$$

This is the basic formula for the δN formalism. A view of the initial and final hypersurfaces is presented in Fig.2.1.

An intriguing point is its independence from the choice of the initial epoch t_i. Thus one can choose the initial flat time slice at any convenient epoch before t_f as long as the perturbation length scale of interest is well outside the Hubble horizon.

The above formula, however, is meaningless unless we have a way to evaluate δN to begin with, and it is almost useless unless it is much simpler than to solve the perturbation equations for $\mathcal{R}_c$. Here the most important property of δN comes into play. It turns out that it can be evaluated solely by solving the background equations, provided that one is given the initial data of the perturbation on the initial flat slice.

Multi-component slow-roll inflation

To be a bit more precise and specific, let us consider slow-roll inflation with a multi-component inflaton, say ϕ^A $(A = 1, 2, 3, \cdots, \mathcal{M})$. The evolution of a homogeneous and isotropic universe in this model is uniquely specified by the values of ϕ^A at a given epoch. In other words, a single trajectory in the $\mathcal{M}$-dimensional field space corresponds to the entire history of a universe. Thus, given a family of trajectories (which never cross each other by definition), one can parametrize each trajectory by N, the number of e-folds, and different trajectories may be labeled by $\mathcal{M}-1$ labels λ^a $(a = 1,2,3,\cdots,\mathcal{M}-1)$. In other words, the evolution of a family of universes is completely determined by specifying the initial condition $\phi^A = \phi^A(\lambda^a)$, which forms an $(\mathcal{M}-1)$-dimensional slice in the field space. Then the number of e-folds from $t = t_i$ to t_f will be a function of $\phi^A(\lambda^a)$,

$$N\left(t_i, t_f\right) = N\left(\phi^A\left(\lambda^a\right)\right). \tag{2.46}$$

Now, let us assume that the background universe is given by a set of initial data $\phi^A = \phi^A(0) \equiv \phi_0^A$,

$$N_0 = N\left(\phi_0^A\right), \tag{2.47}$$

and the perturbation on the initial flat slice at $t = t_i$ is given by $\delta\phi^A(x^i)$. Then one can identify $\phi^A(\lambda^a)$ with $\phi_0^A + \delta\phi^A(x^i)$ by the mapping,

$$x^i \rightarrow \lambda^a\left(x^i\right). \tag{2.48}$$

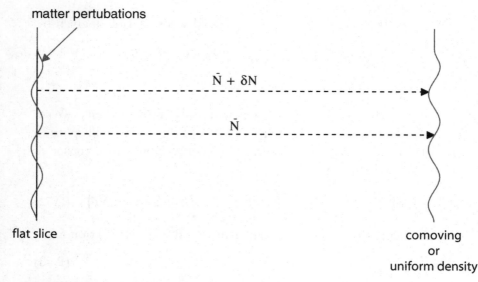

Fig 2.1 An illustration showing the initial and final hypersurfaces for the δN formalism.

Namely,

$$\phi^A (\lambda^a) = \phi_0^A + \delta \phi^A (x^i) . \tag{2.49}$$

Then the difference in the *e*-folding number at linear order is given by

$$\delta N = N \left(\phi^A (\lambda^a) \right) - N \left(\phi_0^A \right) = N \left(\phi_0^A + \delta \phi^A (x^i) \right) - N \left(\phi_0^A \right)$$
$$= \frac{\partial N}{\partial \phi^A} \delta \phi^A (x^i) , \tag{2.50}$$

where the summation over the repeated indices A is understood. It turns out that thus evaluated δN exactly coincides with δN defined in Eq. (2.45).

The above result suggests that if the δN formalism is valid at linear order it may be valid to full non-linear order because we did not use the knowledge of the perturbation equations nor assume the validity of perturbative expansion. The only thing we did was to expand the left-hand side of the above equation to first order in $\delta \phi^A$ at the very last step. In fact in the following subsections, we explicitly verify this and show that the δN formalism can indeed be straightforwardly extended to full non-linear order.

We consider a $\mathcal{M}$ component scalar field whose action is given by

$$S = - \int d^4 x \sqrt{-g} \left[\frac{1}{2} g^{\mu\nu} h_{IJ} \partial_\mu \phi^I \partial_\nu \phi^J + V \left(\phi^K \right) \right]$$
$$(I, J, K = 1, 2, \cdots, \mathcal{M}) \tag{2.51}$$

where h_{IJ} is the metric in the scalar-field space. For simplicity, we set $h_{IJ} = \delta_{IJ}$ below, though the generalization of the scalar-field space to the curved one is straightforward. The energy momentum tensor of the scalar field is given by

$$T_\nu^\mu = \delta_{IJ} g^{\mu\lambda} \partial_\lambda \phi^I \partial_\nu \phi^J - \delta_\nu^\mu \left(\frac{1}{2} g^{\alpha\beta} \delta_{IJ} \partial_\alpha \phi^I \partial_\beta \phi^J + V \left(\phi^K \right) \right) . \tag{2.52}$$

Also we define the background scalar field and its perturbation as

$$\phi^I = \phi_0^I (\eta) + \delta \phi^I . \tag{2.53}$$

For simplicity, we omit the subscript "0" from the background scalar field and denote it by $\phi^I \colon \phi_0^I \to \phi^I$.

When we consider the evolution of the scalar field, it may be more convenient to use the e-folding number as the time coordinate, $dN = Hdt = Hd\eta$. The energy momentum tensor for this scalar field is represented with the time coordinate N,

$$T^{\bar{0}}{}_{\bar{0}} = -(\rho + \delta\rho)$$

$$= -\frac{1}{2}\frac{\mathcal{H}^2}{a^2}\delta_{IJ}\phi^I_N\phi^J_N + V - \frac{\mathcal{H}^2}{a^2}\delta_{IJ}\left(\delta\phi^I_N - \phi^I_N A\right)\phi^J_N - V_I\delta\phi^I,$$

(2.54)

$$T^{\bar{0}}{}_i = -\frac{\mathcal{H}}{a^2}\delta_{IJ}\phi^I_N\partial_i\delta\phi^J,$$

(2.55)

$$T^i{}_j = (P + \delta P)\delta^i{}_j$$

$$= \left[\frac{1}{2}\frac{\mathcal{H}^2}{a^2}\delta_{IJ}\phi^I_N\phi^J_N - V + \frac{\mathcal{H}^2}{a^2}\delta_{IJ}\left(\delta\phi^I_N - \phi^I_N A\right)\phi^J_N - V_I\delta\phi^I\right]\delta^i{}_j,$$

(2.56)

where the barred indices $(\bar{0})$ are used to indicate that N is the time coordinate, the subscript N means a derivative with respect to N. For the sake of brevity, in the remainder of the book, we omit the bar over the time indices. Moreover, V_I and V^I are defined by

$$V_I \equiv \frac{\partial V}{\partial\phi^I}, \quad V^I \equiv \delta^{IJ}V_J.$$

(2.57)

For the background, we have only two independent components in the Einstein equation,

$$3M_P^2\mathcal{H}^2 = a^2\rho = \frac{1}{2}\mathcal{H}^2\delta_{IJ}\phi^I_N\phi^J_N + a^2V,$$

(2.58)

$$M_P^2\left(2\mathcal{H}\mathcal{H}_N + \mathcal{H}^2\right) = -a^2P = -\frac{1}{2}\mathcal{H}^2\delta_{IJ}\phi^I_N\phi^J_N + a^2V,$$

(2.59)

where the former is the $(0,0)$-component (the Friedmann equation) and the latter is the trace part of (i,j)-components. The latter can also be derived by using the Friedmann equation and the scalar field equation. The background scalar field equation is given by

$$\phi^{I\prime\prime} + 2H\phi^{I\prime} + a^2V^I = 0,$$

(2.60)

which may be rewritten with N as the time coordinate,

$$\mathcal{H}\frac{d}{dN}\left(\mathcal{H}\phi_N^I\right) + 2\mathcal{H}^2\phi_N^I + a^2 V^I = 0. \tag{2.61}$$

The linearized Einstein equation has four independent components regarding the scalar perturbations; the $(0,0)$-component, the $(0,i)$-component, and the trace and traceless parts of the (i,j)-components. Using N as the time coordinate, the $(0,0)$-component is

$$6\left(A - R_N\right) + \frac{2}{\mathcal{H}^2}\left[\triangle R - H\left(-\triangle\right)^{1/2}\sigma_g\right]$$
$$= \frac{1}{M_P^2}\left(-\delta_{IJ}\left(\delta\phi_N^I - \phi_N^I A\right)\phi_N^J - \frac{a^2}{\mathcal{H}^2}V_I\delta\phi^I\right), \tag{2.62}$$

the $(0,i)$-component is

$$2\left(A - \mathcal{R}_N\right) = \frac{1}{M_P^2}\delta_{IJ}\phi_N^I\delta\phi^J, \tag{2.63}$$

and the trace and traceless parts of the (i,j)-components are, respectively,

$$2A_N + 2\left(2\frac{\mathcal{H}_N}{\mathcal{H}} + 1\right)A - 2\mathcal{R}_{NN} - 2\left(\frac{\mathcal{H}_N}{\mathcal{H}} + 2\right)\mathcal{R}_N$$
$$+ \frac{2}{3\mathcal{H}^2}\left[\triangle\left(A + \mathcal{R}\right) - \mathcal{H}(-\triangle)^{1/2}\left(\sigma_{gN} + 2\sigma_g\right)\right] \tag{2.64}$$
$$= \frac{1}{M_P^2}\left(\delta_{IJ}\left(\delta\phi_N^I - \phi_N^I A\right)\phi_N^J - \frac{a^2}{\mathcal{H}^2}V_I\delta\phi^I\right),$$

and

$$\triangle\left(A + \mathcal{R}\right) - \mathcal{H}\left(-\triangle\right)^{1/2}\left(\sigma_{gN} + 2\sigma_g\right) = 0, \tag{2.65}$$

where σ_g has been defined in Eq. (2.3). In addition, we have the perturbed scalar field equation,

$$\mathcal{H}^2\delta\phi_{NN}^I + \mathcal{H}\left(\mathcal{H}_N + 2\mathcal{H}\right)\delta\phi_N^I - \triangle\delta\phi^I + a^2 V^I{}_{,J}\delta\phi^J$$
$$+ 2a^2 V^I A - \mathcal{H}^2\phi_N^I A_N + \mathcal{H}\phi_N^I\left[3\mathcal{H}\mathcal{R}_N + (-\triangle)^{1/2}\sigma_g\right] = 0. \tag{2.66}$$

As we have discussed in Sec. 2.1.1, one of the most important quantities in the linear theory is the comoving curvature perturbation $\mathcal{R}_c$, which is the curvature perturbation on the comoving slice $\delta T_i^0 = 0$. In the present case, it is expressed in a gauge-invariant manner as

$$\mathcal{R}_c = \mathcal{R} - \mathcal{H}\frac{\delta_{IJ}\phi^{I'}\delta\phi^J}{\delta_{IJ}\phi^{I'}\phi^{J'}} = \mathcal{R} - \frac{\delta_{IJ}\phi_N^I\delta\phi^J}{\delta_{IJ}\phi_N^I\phi_N^J}. \tag{2.67}$$

From this expression, the relation between $\mathcal{R}_c$ and the perturbation of the scalar field on the flat slicing $\delta\phi_f^I$, where the flat slicing is defined by $\mathcal{R} = 0$ (see Eq. (2.2)), becomes apparent

$$\mathcal{R}_c = -\frac{\delta_{IJ}\phi_N^I\delta\phi_f^J}{\delta_{IJ}\phi_N^I\phi_N^J}. \tag{2.68}$$

In particular, in the single-field case it reduces to

$$\delta\phi_f = -\phi_N\mathcal{R}_c. \tag{2.69}$$

By choosing the flat slice, we can rewrite Eq. (2.66) in terms of scalar fields only. First, from Eq. (2.63), we can express A in terms of $\delta\phi^I$ as

$$2M_P^2 A_f = \delta_{IJ}\phi_N^I\delta\phi_f^J, \tag{2.70}$$

where the subscript f denotes a quantity evaluated on flat slices. Then from Eq. (2.62), σ_g is expressed in terms of A and $\delta\phi^I$ as

$$\begin{aligned}
-\frac{2M_P^2}{\mathcal{H}}(-\triangle)^{1/2}\sigma_{gf} &= -6M_P^2 A_f - \delta_{IJ}\left(\delta\phi_{f,N}^I - \phi_N^I A_f\right)\phi_N^J - \frac{a^2}{\mathcal{H}^2}V_I\delta\phi_f^I \\
&= -\frac{a^2}{\mathcal{H}^2}\left(\frac{V}{M_P^2}\delta_{IJ}\phi_N^J + V_I\right)\delta\phi_f^I - \delta_{IJ}\phi_N^I\delta\phi_{f,N}^J,
\end{aligned} \tag{2.71}$$

where we have used Eq. (2.70) to eliminate A in the second line. Inserting Eq. (2.70) and Eq. (2.71) into Eq. (2.66), we obtain the closed equation for the scalar field,

$$\mathcal{H}^2\delta\phi^I_{f,NN} + \mathcal{H}\left(\mathcal{H}_N + 2\mathcal{H}\right)\delta\phi^I_{f,N} - \triangle\delta\phi^I_f + a^2 V^I{}_J\delta\phi^J_f$$
$$= \frac{1}{M_P^2}\delta_{JK}\left[a^2\frac{V}{M_P^2}\phi^I_N\phi^K_N + \mathcal{H}^2\partial_N\left(\phi^I_N\phi^K_N\right)\right]\delta\phi^J_f, \tag{2.72}$$

which can be rephrased as

$$H^2\delta\phi^I_{f,NN} + H\left(H_N + 3H\right)\delta\phi^I_{f,N} - \frac{\triangle}{a^2}\delta\phi^I_f + V^I{}_J\delta\phi^J_f$$
$$= \frac{1}{M_P^2}\delta_{JK}\left[\frac{V}{M_P^2}\phi^I_N\phi^K_N + H^2\partial_N\left(\phi^I_N\phi^K_N\right)\right]\delta\phi^J_f. \tag{2.73}$$

Under the slow-roll approximation,

$$\epsilon = -\frac{\dot{H}}{H^2} = \left(-\frac{\mathcal{H}_N}{\mathcal{H}} + 1\right) \ll 1, \quad |\eta^I| = \left|\frac{\ddot{\phi}^I}{H\dot{\phi}^I}\right| = \left|\frac{\phi^I_{NN}}{\phi^I_N} - \epsilon\right| \ll 1, \tag{2.74}$$

the background scalar-field equation (2.61) reduces to

$$\phi^I_N \simeq -M_P^2\frac{V^I}{V}; \quad V \simeq 3M_P^2 H^2 = 3M_P^2\frac{\mathcal{H}^2}{a^2}, \tag{2.75}$$

and the perturbed field equation (2.73) can be simplified in the long-wavelength limit, or in the limit where the spatial gradients are neglected, as,

$$3\mathcal{H}^2\delta\phi^I_{f,N} + a^2 V^I{}_J\delta\phi^J_f \simeq \frac{3}{M_P^2}\mathcal{H}^2\phi^I_N\delta_{JK}\phi^J_N\delta\phi^K_f. \tag{2.76}$$

Combining with the background equations, one finally gets

$$\delta\phi^I_{f,N} \simeq \left(-\frac{V^I{}_K}{3H^2} + \frac{\phi^I_N\phi^J_N}{M_P^2}\delta_{JK}\right)\delta\phi^K_f$$
$$\simeq M_P^2\left(-\frac{V^I{}_K}{V} + \frac{V^I V^J\delta_{JK}}{V^2}\right)\delta\phi^K_f. \tag{2.77}$$

2.2.2 δN formalism in multi-field inflation beyond slow-roll

In the previous two subsections we considered the linear δN formalism under the slow-roll approximation. However, it makes the linear δN formalism much more useful if one can formulate it without adopting the slow-roll approximation. In this section, we achieve this purpose in the case of multi-field inflation. We show that on large scales, we can construct the general solution for curvature perturbations from the background solutions by taking the derivative with respect to one of the constants of motion. Using this general solution to the curvature perturbation, we then derive the δN formula with the modification that N is now a function of the phase space of the scalar field, say $N = N(\phi^I, \dot{\phi}^I)$, contrary to the slow-roll case in which N is solely a function of the configuration space, $N = N(\phi^I)$.

N as a time coordinate in uniform $\mathcal{N}$ slicing

First of all, we are only interested in the final conserved curvature perturbation as we have stated before. On superhorizon scales, the curvature perturbation in the adiabatic limit is conserved on the comoving slicing, and it is conserved on any slicings which are equivalent to the comoving slicing at leading order of spatial gradient expansion. Among others, the uniform-energy-density and the uniform-Hubble slicings are the typical ones. However, when the universe is not yet at the adiabatic limit, these time slicings are not necessarily useful on superhorizon scales, because there will be non-trivial evolution of the curvature perturbation whose behavior does not particularly illuminate the physics of the system.

Thus we are to find a more convenient, privileged slicing. Recalling the essence of the δN formula that only the knowledge of the evolution of the homogeneous and isotropic background is necessary for its derivation, we then ask if there exits time slicing in which the perturbation equations in the long-wavelength limit coincide with those governing infinitesimal deviations of the background solution in the homogeneous and isotropic solution space. Namely, the perturbation evolves just like a spatially homogenous perturbation around the background, that is, a small deviation from the background that corresponds to another homogeneous and isotropic universe. If yes, it will be an ideal time slicing since one can readily map the perturbed solution in real space to the derivative vector in the space of homogeneous and isotropic solutions, similar to what we have seen for the slow-roll case in Sec. 2.2.1; see discussions above Eq. (2.45).

In fact as we shall see soon below, such a slicing indeed exists. It is the uniform $\mathcal{N}$ slicing. An important property of this slicing is that it still has a gauge degree of freedom, as it should because the perturbation is equivalent to a spatially homogenous perturbation which always allows a constant shift in the time coordinate. One can then use this freedom to make the initial slice flat.

Our starting point is the gauge transformation rule for $\mathcal{R}$ under an infinitesimal coordinate transformation, $x^\mu \to x^\mu + \Delta x^\mu$. At linear order, $\mathcal{R}$ depends only on the choice of the time slicing, and transforms as

$$\mathcal{R} \to \mathcal{R} - H\Delta t. \tag{2.78}$$

Using N as a time coordinate instead of t, this may be rewritten as

$$\mathcal{R} \to \mathcal{R} - \Delta N. \qquad . \tag{2.79}$$

As given by Eq. (2.40), the number of *e*-folds along the trajectories orthogonal to time slices will be unperturbed if we choose a gauge in which $\mathcal{R}' + (-\triangle)^{1/2}\sigma_g/3 = 0$. This $\mathcal{N} = N$ slicing, which we call the uniform $\mathcal{N}$ slicing, is convenient when we use N as the time coordinate. As mentioned before, the field equations in the long-wavelength limit coincide exactly with those obtained for spatially homogeneous perturbations of the background. In what follows, we therefore adopt the uniform $\mathcal{N}$ slicing and use N as the time coordinate. All the perturbation quantities are those evaluated on this slicing unless otherwise indicated.

The background equations are rewritten as

$$\mathcal{H}\frac{d}{dN}\left(\mathcal{H}\phi_N^I\right) + 2\mathcal{H}^2\phi_N^I + a^2 V^I = 0, \tag{2.80}$$

$$3M_P^2 \mathcal{H}^2 \left(1 - \frac{1}{6}\delta_{IJ}\phi_N^I\phi_N^J\right) = a^2 V, \tag{2.81}$$

where the subscript N means it is a derivative with respect to N. Recall that the second equation is the $(0,0)$-component of the background Einstein equations. The trace part of (i,j)-component gives

$$\frac{1}{H}\frac{dH}{dN} = \frac{1}{\mathcal{H}}\frac{d\mathcal{H}}{dN} - 1 = -\frac{1}{2M_P^2}\delta_{IJ}\phi_N^I\phi_N^J. \tag{2.82}$$

Now, we consider the perturbation in the long-wavelength limit in the uniform N slicing. Neglecting the spatial derivatives, the field equation (2.66) reduces to

$$\mathcal{H}^2\delta\phi_{NN}^I + \left(\mathcal{H}\mathcal{H}_N + 2\mathcal{H}^2\right)\delta\phi_N^I + a^2V^I{}_{,J}\delta\phi^J + 2a^2V^IA - \mathcal{H}^2\phi_N^IA_N = 0. \tag{2.83}$$

We can see that only the perturbation A among the metric perturbations is contained in this equation. Note that A in this gauge may be regarded as the perturbation of the Hubble parameter, as seen from Eq. (2.38). It turns out that A can be expressed in terms of the scalar field perturbation $\delta\phi^I$. From Eq. (2.62), ignoring the term, we have

$$6A = -\frac{1}{M_P^2}\left[\delta_{IJ}\left(\delta\phi_N^I - \phi_N^IA\right)\phi_N^J + \frac{a^2}{\mathcal{H}^2}V_I\delta\phi^I\right], \tag{2.84}$$

which gives

$$a^2VA = -\frac{1}{2}\left(\mathcal{H}^2\delta_{IJ}\delta\phi_N^I\phi_N^J + a^2V_I\delta\phi^I\right). \tag{2.85}$$

After taking the derivative with respect to N of the above equation, we can express A_N by $\delta\phi^I$ and obtain the closed equation for $\delta\phi^I$,

$$\mathcal{H}\frac{d}{dN}\left(\mathcal{H}\delta\phi_N^I\right) + 2\mathcal{H}^2\delta\phi_N^I + a^2\left(V_J^I - \frac{V^IV_J}{V}\right)\delta\phi^J$$
$$- \mathcal{H}^2\left(\frac{1}{M_P^2}\phi_N^I + \frac{V^I}{V}\right)\delta_{JK}\phi_N^K\delta\phi_N^J = 0. \tag{2.86}$$

It is important to note that in order to derive the closed equation for $\delta\phi$, both the $(0,i)$ and the traceless part of the (i,j)-components of the Einstein equations were not needed; they are apparently absent in the background equations. Instead, we only need the $(0,0)$-component and the scalar field equation or the trace part of the spatial components. It is also noteworthy that the trace part of (i, j)-component

$$\mathcal{H}^2 A_N + \mathcal{H}\left(2\mathcal{H}_N + \mathcal{H}\right) A = \frac{1}{2M_P^2}\left[\delta_{IJ}\mathcal{H}^2\left(\delta\phi_N^I - \phi_N^I A\right)\phi_N^J - a^2 V_I \delta\phi^I\right],$$

$$\text{(2.87)}$$

may be solved for A_N by eliminating A with Eq. (2.85),

$$A_N = \frac{1}{M_P^2}\delta_{IJ}\delta\phi_N^I \phi_N^J. \tag{2.88}$$

This agrees with what one would obtain by perturbing Eq. (2.82). This indicates that we are on the right track. If we obtain a complete set of solutions of Eq. (2.86), the only remaining task is to solve for $\mathcal{R}$ and $(-\triangle)^{1/2}\sigma_g$. Let us denote the general solution of the scalar field by

$$\delta\phi^I = c^{(\alpha)}\,\delta\phi_{(\alpha)}^I \tag{2.89}$$

where α runs from 1 to $2\mathcal{M}$, $\delta\phi_{(\alpha)}^I$ are the $2\mathcal{M}$ independent solutions and $c^{(\alpha)}$ are arbitrary constants.

Fortunately one can readily carry out this task because the equation for σ_g is rather simple and also the current gauge condition relates σ_g with $\mathcal{R}'$ in a simple manner. In practice, Eq. (2.65) indicates the rapid decay of σ_g on large scales:

$$(-\triangle)^{1/2}\sigma_g \propto a^{-2}, \tag{2.90}$$

and the condition of the uniform $\mathcal{N}$ slicing, $\mathcal{R}' + (-\triangle)^{1/2}\sigma_g/3 = 0$ also implies the constancy of $\mathcal{R}$ on large scales:

$$\mathcal{R}' \propto (-\triangle)^{1/2}\sigma_g \propto a^{-2} \quad \rightarrow \quad \mathcal{R} = c_R. \tag{2.91}$$

The arbitrariness of the value of $\mathcal{R}$ arose from the remaining gauge degree of freedom in the present gauge condition $\mathcal{R}' = -(-\triangle)^{1/2}\sigma_g/3$. In fact, under an infinitesimal transformation of the time coordinate,

$$N \rightarrow N + \Delta N, \tag{2.92}$$

$\mathcal{R}$ and σ_g transform as

$$\mathcal{R} \to \mathcal{R} - \Delta N \,, \tag{2.93}$$

$$(-\triangle)^{1/2}\sigma_g \to (-\triangle)^{1/2}\sigma_g + \frac{1}{\mathcal{H}}(-\triangle)\,\Delta N. \tag{2.94}$$

Hence for a constant ΔN, the present gauge condition is conserved. Thus this residual gauge degree of freedom corresponds to a constant time translation mode. And now we have $2\mathcal{M}+1$ integration constants for the whole set of perturbation equations in this gauge.

On the other hand, we can easily find a trivial solution

$$\delta \phi = 0 \,, \qquad \mathcal{R} = \text{const.} \tag{2.95}$$

from the $(0,i)$-component of the Einstein equations (2.63) with Eq. (2.85)

$$\mathcal{R}_N = A - \frac{1}{2M_P^2}\delta_{IJ}\phi_N^I \delta\phi^J = \frac{\mathcal{H}^2}{2a^2V}\delta_{IJ}\left(\phi_{NN}^I\delta\phi^J - \phi_N^I\delta\phi_N^J\right). \tag{2.96}$$

Applying the above time translation $N \to N + c$ to the null perturbation, we obtain a pure gauge mode,

$$\delta\phi^I = c\phi_N^I \,, \qquad \mathcal{R} = c \,. \tag{2.97}$$

This indicates that one of the solutions of Eq. (2.86) must be proportional to this time translation mode. In fact it is easy to check that this time translation mode becomes a solution of Eq. (2.86). Then, without loss of generality, we can set

$$\delta\phi_{(1)}^I = \phi_N^I \,. \tag{2.98}$$

One way to investigate the physical degrees of freedom in a system is to construct gauge invariant quantities. A well-known useful quantity in the linear theory is the perturbation of the scalar field on the flat slicing, $\delta\phi_f^I$, defined by the condition $\mathcal{R} = 0$ on that time slicings. In a gauge-invariant manner, it is expressed as

$$\delta\phi_f^I = \delta\phi^I - \phi_N^I \mathcal{R} \,. \tag{2.99}$$

Using the set of $2\mathcal{M}$ independent solutions, the general solution for $\delta\phi_f$ is expressed as

$$\delta\phi_f^I = c^{(\alpha)}\,\delta\phi_{(\alpha)}^I - \phi_N^I c_{\mathcal{R}} \to c^{(\alpha)}\,\delta\phi_{(\alpha)}^I. \qquad (2.100)$$

We note that since the arbitrariness of c_R can be absorbed in the re-definition of $c^{(1)}$, there are only the $2\mathcal{M}$ independent (not $2\mathcal{M} + 1$) integration constants. It is useful to mention that $c^{(1)}$ gives the amplitude of the adiabatic growing mode. Inserting this solution into Eq. (2.68), we see it gives $\mathcal{R}_c = c^{(1)}$.

Now let us reconsider the relation between the solution to Eq. (2.86) and the homogeneous perturbation of the background scalar field equation for a better understanding. Let us assume that we have the general solution for the background equations (2.80) and (2.81). The general solution can be labeled by $2\mathcal{M}$ integration constants. One of them corresponds to the trivial time translation. Thus we may label the general solution by λ^α: $\phi^I(N, \lambda^\alpha)$ $(\alpha = 1, 2, \cdots, 2\mathcal{M} - 1)$. Taking the derivative of Eqs. (2.80) and (2.81) with respect to λ^α, we obtain

$$\mathcal{H}\frac{d}{dN}\left(\mathcal{H}\frac{d}{dN}\phi_\lambda^I\right) + 2\mathcal{H}^2\frac{d}{dN}\phi_\lambda^I$$
$$+ a^2 V^I{}_J\phi_\lambda^J - 2a^2 V^I\frac{\mathcal{H}_\lambda}{\mathcal{H}} + \mathcal{H}^2\phi_N^I\frac{d}{dN}\left(\frac{\mathcal{H}_\lambda}{\mathcal{H}}\right) = 0, \qquad (2.101)$$

$$2a^2 V\frac{\mathcal{H}_\lambda}{\mathcal{H}} - M_P^2\mathcal{H}^2\delta_{IJ}\phi_N^I\frac{d}{dN}\phi_\lambda^J = a^2 V_I\phi_\lambda^I, \qquad (2.102)$$

where the subscript λ means $\partial/\partial\lambda^\alpha$. One can easily see the correspondence between the above equations and Eqs. (2.83) and (2.85). Namely, we can identify

$$\phi_\lambda \Leftrightarrow \delta\phi, \qquad \mathcal{H}_\lambda/\mathcal{H} \Leftrightarrow -A.$$

Thus we conclude that a complete set of solutions on large scales $\delta\phi_{(\alpha)}$ is constructed from the solutions of the background equations. Although this identification is shown only at the linear order, it is not difficult to imagine that it should hold to full non-linear order [30].

Before closing this subsection, it should be noticed that in general the commutativity of the partial derivatives $\partial^2/\partial\lambda^\alpha\partial\lambda^\beta = \partial^2/\partial\lambda^\beta\partial\lambda^\alpha$ is not guaranteed. However under the current gauge condition, since the number of e-folds is not perturbed, the above commutativity is guaranteed. This may be regarded as one of the most important properties of the present gauge condition. There is also another remarkable advantage of this gauge choice which is related to the role of momentum constraint. As discussed in detail in the Appendix of Ref. [17], when we take the derivative of Eqs. (2.80) and (2.81) with respect to λ^α and compare with Eqs. (2.83) and (2.85), the momentum constraint which is absent in background equations plays an important role in general. In other words, without the use of momentum constraint or the essential non-trivial knowledge of perturbations, we cannot verify the coincidence of equations. However, in the uniform $\mathcal{N}$ gauge and also synchronous gauge, one can verify the relation of equations without using momentum constraint. This is possible because of the presence of remaining gauge degree of freedom in these two gauges. By exploiting this remaining degree of freedom, it is always possible to satisfy the momentum constraint. On the other hand, such a treatment cannot be allowed in a general (completely fixed) gauge, since all the gauge degree of freedom is already fixed.

Comoving curvature perturbation

In this subsection, we compute the comoving curvature perturbation on superhorizon scales in a general case, namely for a general potential. Then we explicitly show that the resulting expression for the comoving curvature perturbation in the limit of slow-roll inflation exactly coincides with the previous δN formula, provided that the adiabatic limit is reached at the end of inflation.

The comoving hypersurface is defined by the condition $T^0_{\ i} = 0$. In the case of a scalar field, the $(0, i)$-component of the energy momentum tensor is given by

$$T^0_{\ i} = -\frac{\mathcal{H}}{a^2}\delta_{IJ}\phi_N^I\partial_i\delta\phi^J. \tag{2.103}$$

In the linear theory, this can be rewritten as a total derivative,

$$T^0_{\ i} = -\partial_i\left(\frac{\mathcal{H}}{a^2}\delta_{IJ}\phi_N^I\delta\phi^J\right). \tag{2.104}$$

Then, we can take "comoving slicing" by choosing the time slices in such a way that the inner product of ϕ_N^I and $\delta\phi^I$ vanishes.

In passing, we comment on the comoving slicing in the non-linear case. In the single-scalar-field case we can always take comoving slicing to full non-linear order by taking the uniform scalar field slicing $\phi = \phi(t)$,

$$T^0{}_i = -\frac{\mathcal{H}}{a^2}\phi_N\partial_i\phi(t) = 0. \tag{2.105}$$

However, in the multi-scalar-field case, we cannot generally take the comoving slicing at the non-linear level because T_i^0 cannot be expressed as a total derivative in general. Namely,

$$\partial_j T^0{}_i \neq \partial_i T^0{}_j \quad \text{for} \quad T^0{}_i = -\frac{\mathcal{H}}{a^2}\delta_{IJ}\phi_N^I\partial_i\phi^J. \tag{2.106}$$

Nevertheless, in the linear theory, the comoving curvature perturbation is defined in the gauge-invariant manner in Eq. (2.67), that is,

$$\mathcal{R}_c = \mathcal{R} - \frac{\delta_{IJ}\phi_N^I\delta\phi^J}{\delta_{IJ}\phi_N^I\phi_N^J} = -\frac{\delta_{IJ}\phi_N^I\delta\phi_f^J}{\delta_{IJ}\phi_N^I\phi_N^J}, \tag{2.107}$$

where $\delta\phi_f$ is given by Eq. (2.100). Therefore our task is to determine the coefficients $c^{(\alpha)}$.

In the inflationary universe, quantum fluctuations fuel classical perturbations. Then, in order to give the prediction for the curvature perturbation, we need to quantize the scalar field perturbation. This is usually done for the scalar field perturbation on the flat slicing $\delta\phi_f$ because the equations for $\delta\phi_f$ are closed without non-local terms and the only correction due to the coupling with the metric perturbation appears as a correction to the mass term (mass matrix) of the scalar field.

Therefore, we may assume that $\delta\phi_f$ is given at some initial epoch, $N = N_i$ when the scale of interest has just become superhorizon. Once the initial conditions of $\delta\phi_f^I$ and its time derivative $\partial_N\delta\phi_f^I$ are given, we insert them into Eq. (2.100) to find

$$c^{(\alpha)} = (Q^{-1})^{(\alpha)}{}_J \left(\begin{array}{c} \delta\phi_f^J \\ \delta\phi_{f,N}^J \end{array} \right)\Bigg|_{N=N_i}, \tag{2.108}$$

where Q^{-1} is the inverse matrix of Q defined by

$$Q^J_{(\alpha)} = \begin{pmatrix} \delta\phi^J_{(\alpha)} \\ \delta\phi^J_{(\alpha),N} \end{pmatrix}. \tag{2.109}$$

So far, our treatment has been very general. Now we take the slow-roll limit and check whether Eq. (2.50) obtained under the slow-roll approximation is recovered. We assume that the slow-roll conditions are satisfied for all components of the scalar field,

$$\delta_{IJ}\phi^I_N\phi^J_N \ll 1, \qquad |\phi^I_{NN}| \ll |\phi^I_N|. \tag{2.110}$$

Under this assumption, the background equations reduce to

$$3M_P^2\mathcal{H}^2 = a^2V, \qquad \frac{1}{M_P^2}\phi^I_N = -\frac{a^2V^I}{3\mathcal{H}^2} = -(\ln V)^I, \tag{2.111}$$

where $(\ln V)^I = \delta^{IJ}\partial_J \ln V$. Since the system is described by a first order differential equation, the number of integration constants is reduced to $\mathcal{M}$. The perturbed equation for the scalar field in slow-roll regime is

$$\frac{d}{dN}\phi^I_\lambda = -M_P^2(\ln V)^I_J\phi^J_\lambda. \tag{2.112}$$

Then $\delta\phi^I_f$ is given in terms of the background solution as

$$\delta\phi^I_f = c^{(\alpha)}\delta\phi^I_{(\alpha)} = c^{(\alpha)}\frac{\partial\phi^I}{\partial\lambda^\alpha}, \tag{2.113}$$

where α now runs from 1 to $\mathcal{M}$ because of the slow-roll condition. For convenience, we set $\lambda^\alpha = (N, \lambda^a)$ where $a = 2, 3, \cdots, \mathcal{M}$. Inserting the above equation into Eq. (2.108), we obtain

$$c^{(\alpha)} = (Q^{-1})^\alpha_I\,\delta\phi^I_f = \left[\frac{\partial\lambda^\alpha}{\partial\phi^I}\delta\phi^I_f\right]_{N=N_i}, \tag{2.114}$$

where we have used the fact that, under the slow-roll approximation, one has $Q^I_\alpha = \partial_\alpha\phi^I$. Thus the comoving curvature perturbation $\mathcal{R}_c$ is given by

$$
\begin{aligned}
\mathcal{R}_c &= -c^{(\alpha)} \frac{\delta_{IJ} \phi_N^I \delta\phi_{(\alpha)}^J}{\delta_{IJ} \phi_N^I \phi_N^J} \\
&= -\left(\left[\frac{\partial N}{\partial \phi^I} \delta\phi_f^I \right]_{N=N_i} + \left[\frac{\partial \lambda^a}{\partial \phi^I} \delta\phi_f^I \right]_{N=N_i} \frac{\delta_{IJ} \phi_N^I \phi_{\lambda^a}^J}{\delta_{IJ} \phi_N^I \phi_N^J} \right).
\end{aligned} \tag{2.115}
$$

To compare the above result with the δN formula, we assume that the universe has reached the adiabatic limit at the end of inflation where we set $N = N_e$ for all the background trajectories. Since the adiabatic limit is reached at the end of inflation, $N = N_e$ should be an equal potential surface in the field space. Otherwise, the evolution after inflation would not be unique since different potential energies mean different energy densities, hence different initial conditions for different regions of the universe. This means $\partial V / \partial \lambda^a = 0$ at $N = N_e$. Then from Eq. (2.111), it follows that

$$
0 = \frac{\partial \ln V}{\partial \lambda^a} = \frac{\partial \ln V}{\partial \phi^J} \frac{\partial \phi^J}{\partial \lambda^a} = -\delta_{IJ} \phi_N^I \phi_{\lambda^a}^J. \tag{2.116}
$$

Thus the last term in Eq. (2.115) vanishes at $N = N_e$ to give

$$
\mathcal{R}_c(N_e) = -\left[\frac{\partial N}{\partial \phi^I} \delta\phi_f^I \right]_{N=N_i}. \tag{2.117}
$$

This formula coincides with Eq. (2.50) except for the sign. The sign difference is simply due to the different definition of the number of e-folds, N. Namely, N in this subsection is the time coordinate which flows forward in time, while N in the δN formula is the number of e-folds counted *backward* in time from a given epoch after the universe has reached the adiabatic limit. If we denote the number of e-folds in the δN formula by $N_{\delta N}$, it is related to N in this subsection as $N_{\delta N} = N_e - N$. Thus we have successfully recovered the δN formula.

Towards a non-linear extension

For the purpose of extending the result of the previous subsection to the non-linear case, it is worthwhile to re-derive that result by a gauge transformation between two $\mathcal{N}$ slicings with different values of

$\mathcal{R}$ because it can be generalized to non-linear order. As σ_g decays rapidly on superhorizon scales (Eq. (2.90)), ignoring it in the gauge condition $\mathcal{R}' + (-\triangle)^{1/2}\sigma_g/3 = 0$ implies $\mathcal{R}' = 0$. Thus in particular, one may choose the flat slicing, $\mathcal{R} = 0$. Let us denote $\mathcal{N}$ in the flat slicing by $\mathcal{N}_f$. Another slicing one may choose is the slicing on which $\mathcal{R}$ is constant and coincides with the comoving slice at the end of inflation, i.e., $\mathcal{R}(t) = \mathcal{R}_c(t_e) = \text{const}$.

In the above we have re-introduced the cosmic time t in order not to confuse N as the time coordinate with $\mathcal{N}\,(=N)$ as the number of e-folds between two slices. We first recall the definition of the number of e-folds,

$$\mathcal{N}(t_1, t_2) = \frac{1}{3}\int_{\tau_1}^{\tau_2} K\, d\tau. \qquad (2.118)$$

Taking the flat slicing, we have

$$\mathcal{N}_f(t_i, t_e) = N(t_i, t_e), \quad \mathcal{R}(t_i) = \mathcal{R}(t_e) = 0. \qquad (2.119)$$

Now, we consider an infinitesimal gauge transformation of the time coordinate from the flat slicing to the comoving slicing at $t = t_e$, $t \to \bar{t} = t + \delta t$. This induces the gauge transformation,

$$\mathcal{N}_f \to \mathcal{N}(t_1, t_e - \delta t) = \mathcal{N}_f(t_1, t_e) - \delta N; \quad \delta N = H\delta t(t_e), \qquad (2.120)$$

and

$$\mathcal{R} = 0 \to \mathcal{R}_* = \mathcal{R}_c = 0 - \delta N. \qquad (2.121)$$

Thus we find

$$\mathcal{R}_c(t_e) = -\delta N = \mathcal{N}(t_1, t_e - \delta t) - N(t_1, t_e), \qquad (2.122)$$

where $\mathcal{N}(t_1, t_e - \delta t)$ is the number of e-folds from the flat slice at $t = t_1$ to the comoving slice at $t = t_e$. Below we shall see that there is a natural extension of the curvature perturbation to the non-linear case, with which one can formulate the non-linear δN formula.

2.3   Non-linear δN formalism

In this section, we shall extend the δN formalism to non-linear order. First we will define metric variables based on ADM decomposition and derive the basic equations in terms of those variables. Next we will introduce gradient expansion method as an alternative to standard cosmological perturbation approach. Then applying gradient expansion to the Einstein equations, we again find that the leading order equations under a certain gauge choice, namely the (non-linearly extended) uniform $\mathcal{N}$ gauge, resemble those of background. And then after appropriately extending the definition of curvature perturbation to non-linear order, we shall derive the non-linear δN formula.

2.3.1   *The Einstein equations*

We develop a theory of non-linear cosmological perturbations on superhorizon scales. For this purpose we employ the ADM formalism and the gradient expansion approach which will be explained in detail in the next subsection. In the ADM decomposition, the metric is expressed as

$$ds^2 = g_{\mu\nu}dx^\mu dx^\nu = -\alpha^2 dt^2 + \hat{\gamma}_{ij}\left(dx^i + \beta^i dt\right)\left(dx^j + \beta^j dt\right), \quad (2.123)$$

where α is the lapse function, β^i is the shift vector while $\hat{\gamma}_{ij}$ represents the spatial three-metric. Moreover, it should be noted that Latin indices run over $1, 2$ and 3. The extrinsic curvature K_{ij}, (the conjugate momentum of the spatial metric) can be introduced by

$$K_{ij} = \frac{1}{2\alpha}\left(\partial_t \hat{\gamma}_{ij} - \hat{D}_i \beta_j - \hat{D}_j \beta_i\right), \quad (2.124)$$

where $\hat{D}$ is the covariant derivative with respect to the spatial metric $\hat{\gamma}_{ij}$. In addition to the standard ADM decomposition, the spatial metric and the extrinsic curvature are further decomposed so as to separate trace and trace-free parts

$$\hat{\gamma}_{ij} = a^2\left(t\right)e^{2\psi}\gamma_{ij}; \quad \det\gamma_{ij} = 1, \quad (2.125)$$

$$K_{ij} = a^2\left(t\right)e^{2\psi}\left(\frac{1}{3}K\gamma_{ij} + A_{ij}\right); \quad \gamma^{ij}A_{ij} = 0, \quad (2.126)$$

where $a(t)$ is the scale factor of a fiducial homogeneous FLRW space-time and the determinant of γ_{ij} is normalized to be unity and A_{ij} is trace free. The explicit form of K is given by

$$K \equiv \hat{\gamma}^{ij} K_{ij} = \frac{1}{\alpha} \left[3(H + \partial_t \psi) - \hat{D}_i \beta^i \right], \qquad (2.127)$$

where H is the Hubble parameter defined by $H(t) \equiv \dot{a}(t)/a(t)$.

As for the matter field, let us consider the same $\mathcal{M}$-component scalar field, ϕ^I, as in Eq. (4.1). In the ADM decomposition, after contracting with the unit normal vector n^μ such that

$$n_\mu dx^\mu = -\alpha dt, \quad n^\mu \partial_\mu = \frac{1}{\alpha} \left(\partial_t - \beta^i \partial_i \right), \qquad (2.128)$$

all the independent components of the energy momentum tensor are conveniently expressed in terms of E, J_i and T_{ij} as

$$E \equiv T_{\mu\nu} n^\mu n^\nu, \quad J_i \equiv -T_{i\mu} n^\mu, \quad T_{ij} = T_{ij}. \qquad (2.129)$$

For convenience, we further decompose T_{ij} in the same way as Eq. (2.126),

$$T_{ij} = a^2(t) e^{2\psi} \left(\frac{1}{3} S \gamma_{ij} + S_{ij} \right); \quad S \equiv \hat{\gamma}^{ij} T_{ij}, \qquad (2.130)$$

Now we write down the Einstein equations. In the ADM decomposition, the Einstein equations are separated into four constraints, namely the Hamiltonian and three momentum constraints, and six dynamical equations for the spatial metric. The constraints are

$$\frac{1}{a^2 e^{2\psi}} \left[R - (4D^2 \psi + 2D^i \psi D_i \psi) \right] + \frac{2}{3} K^2 - A_{ij} A^{ij} = 2E, \qquad (2.131)$$

$$\frac{2}{3} \partial_i K - e^{-3\psi} D_j \left(e^{3\psi} A_i^j \right) = J_i. \qquad (2.132)$$

Here $R \equiv R[\gamma]$ is the three-dimensional Ricci scalar of the normalized spatial metric γ_{ij} and γ^{ij} is the inverse of γ_{ij}. D_i is the covariant derivative with respect to γ_{ij} and the spatial indices are raised or lowered by γ^{ij} and γ_{ij}, respectively. Finally D^2 represents $\gamma^{ij} D_i D_j$.

As for the dynamical equations of the spatial metric, we rewrite Eq. (2.124) as

$$\partial_\perp \psi = -\frac{H}{\alpha} + \frac{1}{3}\left(K + \frac{\partial_i \beta^i}{\alpha}\right), \tag{2.133}$$

$$\partial_\perp \gamma_{ij} = 2A_{ij} + \frac{1}{\alpha}\left(\gamma_{ik}\partial_j \beta^k + \gamma_{jk}\partial_i \beta^k\right)^{TF}. \tag{2.134}$$

The equations for the extrinsic curvature (K, A_{ij}) are given by

$$\partial_\perp K = -\frac{1}{3}K^2 - A_{ij}A^{ij} + \frac{1}{a^2 e^{2\psi}\alpha}\left(D^2\alpha + D^i\alpha D_i\psi\right) - \frac{1}{2}\left(S + E\right), \tag{2.135}$$

$$\partial_\perp A_{ij} = -KA_{ij} + 2A_i^k A_{kj} + \frac{1}{\alpha}\left(A_{ik}\partial_j\beta^k + A_{jk}\partial_i\beta^k - \frac{2}{3}A_{ij}\partial_k\beta^k\right)$$

$$-\frac{1}{a^2 e^{2\psi}}\left[R_{ij} + D_i\psi D_j\psi - D_i D_j\psi - \frac{1}{\alpha}\left(D_i D_j\alpha - D_i\alpha D_j\psi\right.\right.$$

$$\left.\left.-D_j\psi D_i\alpha\right)\right]^{TF} + S_{ij}, \tag{2.136}$$

where $\partial_\perp \equiv n^\mu \partial_\mu$, and we have introduced the trace-free projection operator $[...]^{TF}$ defined for a tensor Q_{ij} as

$$Q_{ij}^{TF} \equiv Q_{ij} - \frac{1}{3}\gamma_{ij}\gamma^{kl}Q_{kl}. \tag{2.137}$$

Finally, the equation of motion for the scalar field is

$$\partial_\perp\left(\partial_\perp\phi^I\right) + K\partial_\perp\phi^I - \frac{1}{\alpha a^3 e^{3\psi}}D_i\left(\alpha a e^\psi \gamma^{ij}\partial_j\phi^I\right) + V^I = 0. \tag{2.138}$$

2.3.2   *Gradient expansion*

In cosmological perturbation theory we expand the field equations in powers of the perturbations. Specifically we expand the equations in powers of a small parameter, say δ, which characterizes the smallness of perturbations. In particular, the linear perturbation theory truncates the equations at linear order in δ.

Since we are to study nonlinear perturbations on superhorizon scales, the above standard perturbative approach will not be very useful. On the other hand, the spatial derivative of the metric, say, will be suppressed by a factor H^{-1}/λ ($\ll 1$) relative to the time derivative on superhorizon scales, where λ is the characteristic length scale of the perturbation under consideration. In this case, we may employ the method of spatial gradient expansion [20,31–34], which is an expansion in the spatial gradient of these perturbations. Namely, we attach a parameter ε to each spatial gradient and expand the equations in powers of ε. We call it gradient expansion by omitting the adjective "spatial" for simplicity. The equations at leading non-trivial order in gradient expansion are obtained by truncating the exact equations at first order in ε and set $\varepsilon = 1$ at the end.

The gradient expansion is useful when every quantity can be assumed to be sufficiently smooth on scales greater than the Hubble horizon scale, $H^{-1} = (\dot{a}/a)^{-1}$. In other words, we consider a universe that looks locally homogeneous and isotropic on scales smaller than λ_* where $\lambda_* \gg H^{-1}$.[1] Without going to technical details, we assume that this is the case for the actual universe, with some reasonable coarse-graining or smoothing procedure.

In the perturbative approach, since any physical perturbation on a spatially homogeneous and isotropic background will have spatial dependence by definition, it will be multiplied by the fictitious parameter ε at least once. Thus the linear perturbation on superhorizon scales should agree with the linearized version of the corresponding perturbation in gradient expansion at leading order. This implies that the results obtained in gradient expansion will closely resemble those of linear perturbation theory as long as superhorizon perturbations are concerned.

To make the relation between the perturbative approach with the gradient expansion approach a little more specific, let us consider the Fourier component of a long wavelength perturbation where the ratio of the comoving wavenumber to the comoving Hubble scale $k/aH = k/\mathcal{H}$ is small. Then the equations can be expanded in powers of k/aH,

---

[1]The cosmic scale factor a may not be well defined on very large scales, since the expansion rate actually depends on the choice of time slicing. Nevertheless, introducing the spatial dependence, $a = a(t, x^i)$, and setting $H \equiv \dot{a}/a$, both the local scale factor and local Hubble parameter become well defined in the limit $|\partial_i a|/a \ll Ha$.

and it corresponds to the expansion in powers of spatial derivatives when inverse-Fourier-transformed to the real space. This implies that we may make the identification,

$$\varepsilon \equiv \frac{k}{aH} = \frac{k}{\mathcal{H}}. \qquad (2.139)$$

Although this correspondence is exact only in the linear perturbation theory, it is useful even for non-linear perturbations when one considers not the equations but the resulting physical quantities as long as one can consistently introduce a fiducial homogeneous background. For example, assuming $k/\mathcal{H} \ll 1$, we may consider an intermediate scale k_*, $\varepsilon = k/\mathcal{H} \ll k_*/\mathcal{H} \ll 1$, where the region of the comoving size k_*^{-1} may be well approximated by a homogeneous and isotropic FLRW universe.

Thus our key physical assumption is that in the limit $\varepsilon \to 0$, corresponding to a sufficiently large smoothing scale, the universe becomes *locally* homogeneous and isotropic. This (i.e. the gradient expansion at leading order) is called the separate universe hypothesis [17,15,35]. Under this assumption, the Hubble horizon size is the only relevant scale to which all the derivatives are to be compared. Of course, this may not be the case in general. For example, one may consider a locally spatially anisotropic universe, in which no region of any scale can be approximated by a homogeneous and isotropic FLRW universe. Nevertheless, since our main interest is in the inflationary cosmology in which a locally homogeneous and isotropic universe is realized under reasonable assumptions (cosmic no-hair conjecture), the separate universe hypothesis is expected to be valid in the actual Universe.

An immediate consequence of our assumption is that the metric should reduce to that of the FLRW locally. Namely, with an appropriate choice of the coordinates, the metric of any local region takes the following form,

$$ds^2 = -dt^2 + a^2(t)\,\delta_{ij}dx^i dx^j, \qquad (2.140)$$

where the spatial metric is chosen to be flat because any non-vanishing spatial curvature may be regarded as a perturbation on this background in the context of spatial gradient expansion.

Let us now investigate the implications of the above form of the metric. In the limit $\varepsilon \to 0$, the above local metric should be globally valid. This implies that the metric component β_i should vanish in this limit as $\beta_i = \mathcal{O}(\varepsilon)$.[2] It may be noted, however, that this is not really a necessary condition but rather a matter of choice of coordinates for convenience.

Now, let us turn to the spatial components γ_{ij}, defined by det $\gamma = 1$. While a homogeneous time-independent γ_{ij} can be locally eliminated by a spatial coordinate transformation but a homogeneous time-dependent γ_{ij} cannot be eliminated by any coordinate transformation. Therefore we must require $\dot{\gamma}_{ij} \sim A_{ij} = \mathcal{O}(\varepsilon)$. In fact, if the cosmic no-hair conjecture is valid in the context of inflationary cosmology, which is true in the case of Einstein gravity, a stronger condition, $\dot{\gamma}_{ij} \sim A_{ij} = \mathcal{O}(\varepsilon^2)$, holds. We note that in alternative theories of gravity, the assumption $\dot{\gamma}_{ij} \sim A_{ij} = \mathcal{O}(\varepsilon^2)$ may not be as natural as in the Einstein case.

The conditions on the metric components which we require are therefore

$$\beta_i = \mathcal{O}(\varepsilon), \tag{2.141}$$

$$\frac{1}{\alpha}\dot{\gamma}_{ij} \sim A_{ij} = \mathcal{O}(\varepsilon^2). \tag{2.142}$$

Note that there is no requirement on ψ and α. In view of Eq. (2.141), the line element simplifies to

$$ds^2 = -\alpha^2 dt^2 + 2\beta_i dx^i dt + a^2(t)\, e^{2\psi}\gamma_{ij}dx^i dx^j. \tag{2.143}$$

2.3.3 *Leading order in gradient expansion*

In this subsection, we study Einstein's field equations to leading order in gradient expansion and make clear the correspondence between the leading order equations and background equations. This correspondence can be used to construct the solution of perturbed equations in terms of the background solution.

---

[2]We adopt the traditional mathematics notation [36], according to which $f = \mathcal{O}(\varepsilon^n)$ means that f falls like ε^n or faster.

Now we derive the Einstein equations, particularly those that are relevant at the leading order in gradient expansion from the original Einstein equations (2.131), (2.132), (2.135), (2.136) and scalar field equation (2.138). First of all, according to the number of spatial derivatives and ordering of β_i and $\dot{\gamma}_{ij}$, Eqs. (2.141), (2.142), one finds to leading order in gradient expansion, Eqs. (2.131) and (2.135) can be written as

$$\frac{1}{3}K^2 = \frac{1}{2}\delta_{IJ}\partial_\tau\phi^I\partial_\tau\phi^J + V + \mathcal{O}\left(\varepsilon^2\right), \qquad (2.144)$$

$$\partial_\tau K = -\frac{3}{2}\delta_{IJ}\partial_\tau\phi^I\partial_\tau\phi^J + \mathcal{O}\left(\varepsilon^2\right). \qquad (2.145)$$

Furthermore the scalar field equation (2.138) yields

$$\partial_\tau\left(\partial_\tau\phi^I\right) + K\partial_\tau\phi^I + V^I + \mathcal{O}\left(\varepsilon^2\right) = 0. \qquad (2.146)$$

Here we have introduced the proper time τ by

$$\tau\left(t_2,t_1;x^i\right) \equiv \int_{t_1}^{t_2} dt\,\alpha\left(t,x^i\right)|_{x^i=\text{const.}}. \qquad (2.147)$$

In terms of τ, the expression of K in Eq. (2.127) is also simplified as

$$K = \frac{1}{\alpha}\frac{\partial_t\left(a^3 e^{3\psi}\right)}{a^3 e^{3\psi}} = 3\frac{\partial_\tau\left(ae^\psi\right)}{ae^\psi} + \mathcal{O}\left(\varepsilon^2\right), \qquad (2.148)$$

using the fact that $\beta_i = \mathcal{O}(\varepsilon)$. On the other hand, one will notice that Eq. (2.136) does not contribute at the leading order in gradient expansion since they are at least of the order of ε^2. Moreover, the Eq. (2.132) which starts from the order of ε reduces to

$$\partial_i K = -\frac{3}{2}\delta_{IJ}\partial_\tau\phi^I\partial_i\phi^J + \mathcal{O}\left(\varepsilon^3\right). \qquad (2.149)$$

Since we have mostly used N instead of t as a time coordinate in the last section, it would also be better to rewrite the above equations in terms of the number of e-folds:

$$\mathcal{N}\left(t_2,t_1;x^i\right) \equiv \frac{1}{3}\int_{\tau_1}^{\tau_2} d\tau\,K\left(t,x^i\right)|_{x^i=\text{const.}}$$

$$= \frac{1}{3}\int_{t_1}^{t_2} dt\,\alpha\left(t,x^i\right)K\left(t,x^i\right)|_{x^i=\text{const.}} \qquad (2.150)$$

where and hereafter we omit $\mathcal{O}(\epsilon^2)$ corrections. Then one obtains

$$\frac{1}{3}K^2 = \frac{1}{18}K^2\delta_{IJ}\partial_{\mathcal{N}}\phi^I\partial_{\mathcal{N}}\phi^J + V\,, \qquad (2.151)$$

$$\partial_i K = -\frac{1}{2}K\delta_{IJ}\partial_{\mathcal{N}}\phi^I\partial_i\phi^J\,, \qquad (2.152)$$

$$K\partial_{\mathcal{N}}K = -\frac{1}{2}K^2\delta_{IJ}\partial_{\mathcal{N}}\phi^I\partial_{\mathcal{N}}\phi^J\,, \qquad (2.153)$$

and

$$\frac{1}{3}K\partial_{\mathcal{N}}\left(\frac{1}{3}K\partial_{\mathcal{N}}\phi^I\right) + \frac{1}{3}K^2\partial_{\mathcal{N}}\phi^I + V^I = 0\,. \qquad (2.154)$$

On the other hand, among the background equations, the Friedmann equation and scalar field equation can be rewritten with N as a time coordinate,

$$3H^2 = \frac{1}{2}H^2\phi_N^2 + V\,, \qquad (2.155)$$

$$H\partial_N(H\partial_N\phi) + 3H^2\partial_N\phi + V_\phi = 0\,. \qquad (2.156)$$

Now one will easily notice that the basic equations at leading order in gradient expansion, Eqs. (2.151) and (2.153), take exactly the same form as those in the background with the identifications:

$$\frac{1}{3}K \Leftrightarrow H, \qquad \mathcal{N} \Leftrightarrow N\,. \qquad (2.157)$$

This implies that the physics on the $\mathcal{N} = $ const. slicings are the same as the background physics since both basic equations are identical. If we choose other slicings, it will be difficult to solve the leading order equations in terms of the background solutions since the basic equations are simply different. In other words, there is a preferred slicing on which the physics can be easily related with the background physics. In fact by choosing the uniform $\mathcal{N}$ slicing, one can easily construct solutions in terms of background solutions. Namely, given a background solution,

$$\phi^I(N)\Big|_{\text{background}} = \phi_{\text{BG}}^I\left[N; \phi_0^J, \partial_N\phi_0^J\right]\,, \qquad (2.158)$$

with the initial condition $(\phi^I, \partial_N \phi) = (\phi_0^I, \partial_N \phi_0^I)$ at $N = N_0$, one can construct the solution at leading order in gradient expansion as

$$\phi^I(\mathcal{N}, x^i)\Big|_{\text{gradient}} = \phi_{\text{BG}}^I\left[\mathcal{N}, \phi_0^J(x^i), \partial_N \phi_0^J(x^i)\right], \qquad (2.159)$$

where ϕ_0^I and $\partial_N \phi_0^I$ now depend on x^i. As is clear from this equation, all the information of inhomogeneities is contained in the initial condition of the scalar field.

2.3.4 *Curvature perturbation and non-linear δN formula*

Here first let us define the non-linear curvature perturbation — to be precise, an appropriate curvature perturbation at leading order in gradient expansion.

To begin with, in a linear theory, a metric component $\mathcal{R}$ in Eq. (2.2) is called a curvature perturbation, because it corresponds to the three-dimensional Ricci scalar,

$$R^{(3)}\Big|^{\text{linear}} = -\frac{4}{a^2} \triangle \mathcal{R}. \qquad (2.160)$$

On the other hand, one can rewrite the line elements of spatial (3D) section in a different manner:

$$ds^2\Big|^{\text{3D}} = a^2(\eta)\left[\left(1 + 2\mathcal{R} + \frac{2}{3}C\right)\delta_{ij} - 2\left((-\triangle)^{-1}C_{,ij} + \frac{1}{3}C\delta_{ij}\right)\right]dx^i dx^j, \qquad (2.161)$$

where the inside of the first parentheses corresponds to the trace part of the spatial metric and the second to the traceless part, or in other words, the unimodular part of the spatial metric. Now the correspondence to the non-linear metric, Eq. (2.123) is clear:

$$\psi \Leftrightarrow \mathcal{R} + \frac{1}{3}C, \qquad \gamma_{ij}\Big|_{\text{scalar}} \Leftrightarrow C. \qquad (2.162)$$

Although both ψ and γ_{ij} can contribute to the Ricci scalar in general, we can safely neglect the contribution from γ_{ij} since there is no essential dynamics in γ_{ij} at leading order in gradient expansion. And hence now one arrives at an appropriate definition of the non-linear curvature perturbation at this order as

$$\mathcal{R}^{NL} \sim \psi + \mathcal{O}\left(\varepsilon^2\right). \tag{2.163}$$

Of course in the limit of linear theory, we recover the conventional definition of linear curvature perturbation since ψ reduces to $\mathcal{R}$ at linear order where C ($\sim$ shear) can be neglected on large scales.

Now, we shall extend the linear δN formalism up to non-linear order. The expression of number of e-folds in Eq. (2.150) can be rewritten with the aid of Eq. (2.148),

$$\mathcal{N}\left(t_2,t_1;x^i\right) = \int_{t_1}^{t_2} dt\,\left(H + \partial_t\psi\right) = N\left(t_2,t_1\right) + \left[\psi\left(t_2,x^i\right) - \psi\left(t_1,x^i\right)\right]. \tag{2.164}$$

From this equation, we can immediately find that $\mathcal{N}$ coincides with N on "flat" slicings, $\psi = 0$.

Let us consider two different slicings, say slicing A and B. We assume that the two slices coincide at an initial time $t = t_{\mathrm{ini}}$. Then, the difference in the curvature perturbations at the final time $t = t_{\mathrm{fin}}$ is expressed in terms of the difference of the e-folding number,

$$\psi_A\left(t_{\mathrm{fin}},x^i\right) - \psi_B\left(t_{\mathrm{fin}},x^i\right) = \mathcal{N}_A\left(t_{\mathrm{fin}},t_{\mathrm{ini}};x^i\right) - \mathcal{N}_B\left(t_{\mathrm{fin}},t_{\mathrm{ini}};x^i\right)$$

$$\equiv \delta\mathcal{N}_{A\leftarrow B}\left(t_{\mathrm{fin}},t_{\mathrm{ini}};x^i\right). \tag{2.165}$$

We choose the slicing A such that it coincides with a flat slicing at $t = t_{\mathrm{ini}}$ and a uniform energy density slicing at $t = t_{\mathrm{fin}}$ and also the slicing B such that it coincides with a flat slicing all the time. Thus, we have the e-folding number,

$$\psi_A\left(t_{\mathrm{fin}},x^i\right) = \mathcal{N}_A\left(t_{\mathrm{fin}},t_{\mathrm{ini}};x^i\right) - N\left(t_{\mathrm{fin}},t_{\mathrm{ini}}\right)$$

$$\equiv \delta\mathcal{N}_A\left(t_{\mathrm{fin}},t_{\mathrm{ini}};x^i\right), \tag{2.166}$$

where $\delta\mathcal{N}_A$ is the difference between the non-linear e-folding number and the background e-folding number, in other words, the uniform energy density slice and the flat slice. This gives the non-linear generalization of the linear δN formula.

To be specific, let us consider the multi-scalar inflation. Let us further assume the slow-roll inflation for simplicity. As discussed before, one can identify $\mathcal{N}$ with N, and in the slow-roll inflation the $\partial_N \phi_0$ dependence in Eq. (2.159) disappears. Hence we have

$$\phi^I(\mathcal{N}, x^i)\Big|_{\text{gradient}} = \phi^I_{\text{BG}}\left[\mathcal{N}, \phi^J_0\left(x^i, N_0\right)\right] . \tag{2.167}$$

At each spatial point x^i, we have a set of solutions specified by the initial condition $\phi^J = \phi^J_0$ at $\mathcal{N} = N_0$. Exactly the same as the case of linear theory, given by Eqs. (2.30) and (2.31), one can then consider the mapping,

$$x^i \to \lambda^a\left(x^i\right), \tag{2.168}$$

which gives

$$\phi^I\left(\mathcal{N}, x^i\right) = \phi^I_{\text{BG}}\left(\mathcal{N}, \lambda^a\right). \tag{2.169}$$

Then the field space spanned by ϕ^I_{BG} may be also spanned by the trajectories of solutions $\phi^I = \phi^I(\lambda^a, N)$, where and in what follows we identify $\mathcal{N}$ with N and omit the subscript BG for notational simplicity. We also assume that there is no degeneracy in the solutions. One can then invert $\phi^I = \phi^I(\lambda^a, N)$ to obtain

$$(N, \lambda^a) = \left(N\left(\phi^I\right), \lambda^a\left(\phi^I\right)\right). \tag{2.170}$$

In particular, we now have a slicing of the field space in terms of the hypersurfaces of constant N,

$$N = N\left(\phi^I\right) \quad \text{with}\, N_0 = N\left(\phi^I_0\right). \tag{2.171}$$

Here it should be noted that the origin of N can be arbitrarily chosen, corresponding to a constant shift of the time coordinate for each trajectory. For a set of trajectories, one can regard this as a freedom in the choice of constant N surfaces in the field space,

$$N = N\left(\phi^I\right) \quad \to \quad N = N\left(\phi^I\right) + \Delta N\left(\lambda^a\left(\phi^I\right)\right). \tag{2.172}$$

Now we assume that the universe has arrived at the adiabatic limit by the time, say $N = N_e$. We may assume N_e to be independent of the trajectories, since all the trajectories have converged to a unique trajectory by that time. If the evolution of the universe may be still described by ϕ^I in the adiabatic limit, this implies that ϕ^I must be independent of λ^a at $N = N_e$,

$$\phi^I(N_e, \lambda^a) = \phi^I(N_e). \qquad (2.173)$$

Apparently this is possible only if we use the freedom of shifting N for each trajectory, which results in shifting N_0 by some amount ΔN_0 that depends on λ^a. Namely, for a given number of e-folds $N_e - N_0 = f(\lambda^a) + \text{const.}$ for each trajectory specified by λ^a, where we may assume $f(0) = 0$ without loss of generality, we shift the value of N_0 as

$$N_{0,\text{old}} \rightarrow N_0 = N_{0,\text{old}} + \Delta N_0; \quad \Delta N_0 = f(\lambda^a). \qquad (2.174)$$

Then $N_e - N_0$ becomes independent of the trajectories, and the number of e-folds counted backward from $N = N_e$ to $N = N_0$ ($N_e > N_0$) will be given by

$$N_e - N_0 = N_e - N(\phi_0^I). \qquad (2.175)$$

Thus for a fixed N_e, independent of the trajectories in the adiabatic limit, the variation of ϕ_0^I in the field space (which corresponds to the spatial variation of ϕ_0^I) gives rise to the variation of the number of e-folds,

$$\begin{aligned} \delta N &= [N_e - N(\phi_0^I + \delta\phi_0^I)] - [N_e - N(\phi_0^I)] \\ &= -N(\phi_0^I + \delta\phi_0^I) - (-N(\phi_0^I)). \end{aligned} \qquad (2.176)$$

This is the δN formula. The minus signs on the right-hand side are due to the fact that the number of e-folds in the δN formulas is the one counted backward in time from a fixed given time after the universe has arrived at the adiabatic limit.

2.4   Statistical quantities

2.4.1   *Power spectrum in cosmological perturbation theory*

To compute the power spectrum of $\mathcal{R}_c$, we need to compute that of $\delta\phi_*$. As we saw in Eq. (2.73), the equation for the scalar field fluctuations on flat slices is given by

$$\ddot{\delta\phi}_{\mathrm{fl}} + 3H\dot{\delta\phi}_{\mathrm{fl}} + \left[\frac{k^2}{a^2} + V_{,\phi\phi} + \frac{2}{M_P^2}\frac{d}{dt}\left(\frac{V}{H}\right)\right]\delta\phi_{\mathrm{fl}} = 0, \qquad (2.177)$$

where the potential term may be expressed in terms of ϵ defined in Eq. (2.21) and η defined in Eq. (2.22) as

$$V_{,\phi\phi} + \frac{2}{M_P^2}\frac{d}{dt}\left(\frac{V}{H}\right) = V_{,\phi\phi} - 2\epsilon\left(3 - \epsilon + \eta\right)H^2. \qquad (2.178)$$

Then under the slow-roll approximation, $\epsilon_V \ll 1$ and $|\eta_V| \ll 1$, this is approximated as

$$V_{,\phi\phi} + \frac{2}{M_P^2}\frac{d}{dt}\left(\frac{V}{H}\right) \approx \left(3\eta_V - 6\epsilon_V\right)H^2, \qquad (2.179)$$

where η_V and ϵ_V are the slow-roll parameters defined in Eq. (2.26). Hence they are completely negligible on subhorizon scales, $k > aH$. Namely, the field may be well approximated as a minimally coupled massless field in a given background universe. Furthermore, for $k > aH$, since ϵ represents the time variation of the Hubble parameter within one Hubble expansion time, the background can be well approximated by the de Sitter universe with a constant H (locally de Sitter in the sense that H may be regarded as a constant in the computation and its slow time dependence may be recovered in the final, obtained formula).

Then the quantization may be easily done. As usual, expanding the field in terms of the comoving wavenumber, we have

$$\delta\phi_{\mathrm{fl}}(t,\mathbf{x}) = \int \frac{d^3k}{(2\pi)^3}\delta\phi_{\mathrm{fl}}(t,\mathbf{k})e^{i\mathbf{k}\cdot\mathbf{x}};$$

$$\delta\phi_{\mathrm{fl}}(t,\mathbf{k}) \equiv (2\pi)^{3/2}\left(a_{\mathbf{k}}\varphi_k(t) + a^\dagger_{-\mathbf{k}}\varphi_k^*(t)\right), \qquad (2.180)$$

where $a_{\mathbf{k}}$ and $a_{\mathbf{k}}^{\dagger}$ are the annihilation and creation operators, respectively, of the mode $\mathbf{k}$, which satisfy

$$\left[a_{\mathbf{k}}, a_{\mathbf{p}}^{\dagger}\right] = \delta^3\left(\mathbf{k} - \mathbf{p}\right), \quad \left[a_{\mathbf{k}}, a_{\mathbf{p}}\right] = \left[a_{\mathbf{k}}^{\dagger}, a_{\mathbf{p}}^{\dagger}\right] = 0, \tag{2.181}$$

and $\varphi_k(t)$ is the positive frequency mode function which satisfies Eq. (2.177) with the Klein–Gordon normalization,

$$\varphi_k \dot{\varphi}_k^* - \dot{\varphi}_k \varphi_k^* = \frac{i}{a^3}, \tag{2.182}$$

which is derived using equal-time commutation relation between $\phi(\mathbf{x})$ and its canonical momentum [37,38]. In the above, we have used the same symbols for both real space and Fourier space variables for simplicity, and will do so in what follows unless confusion may arise. We also note the different normalization in the definition of the Fourier modes in comparison with the standard field-theoretical definition. The above normalization makes the form of the higher order correlations simpler as we shall see below.

Imposing the Minkowski vacuum (or Bunch–Davies vacuum in the de Sitter approximation) on the initial condition when the mode is deep inside the horizon, $k \gg aH$, we obtain at the leading order in slow-roll expansion,

$$\varphi_k(t) = \frac{H}{\sqrt{2k^3}}\left(i + \frac{k}{Ha}\right)e^{-ik\int^t dt/a}. \tag{2.183}$$

As we immediately see from this, φ_k quickly approaches a constant at $k < aH$, $\varphi_k \to H/\sqrt{2k^3}$ apart from an irrelevant phase factor. Thus we obtain the spectrum of $\delta\phi_*$ at $t = t_k$ as

$$P_{\delta\phi_*}(k) = |\varphi_k(t_k)|^2 = \frac{H^2(t_k)}{2k^3}, \tag{2.184}$$

where in the rigorous sense t_k should be regarded as the time a couple of e-folds after the horizon crossing when the decaying contribution has become negligible. Nevertheless, because $\epsilon = -\dot{H}/H^2 \ll 1$, the identification of t_k with the horizon-crossing time $k = a(t_k)H(t_k)$ on the rightmost side of the above equation is valid with good accuracy. The most important implication of the above result is that it is a generic conclusion for any scalar field whose potential satisfies the slow-roll conditions (2.26).

From Eqs. (2.35) and (2.184), the power spectrum of the conserved comoving curvature perturbation is obtained to be

$$
P_{\mathcal{R}_c}(k) = \frac{1}{M_P^4}\left(\frac{V}{V_{,\phi}}\right)^2\bigg|_{t_k}\, P_{\delta\phi_*}(k) = \frac{1}{M_P^4}\left(\frac{V}{V_{,\phi}}\right)^2\frac{H^2}{2k^3}\bigg|_{t_k}. \qquad (2.185)
$$

In terms of the slow-roll parameter ϵ, noting Eq. (2.26), the power spectrum can be cast into the form

$$
P_{\mathcal{R}_c}(k) = \frac{1}{2M_P^2\epsilon}\frac{H^2}{2k^3}\bigg|_{t_k}. \qquad (2.186)
$$

This result, derived by using the δN formalism, is of course in full agreement with the one obtained by the standard perturbation theory computation.

It is often more convenient to put the power spectrum in the dimensionless form, i.e. the power in the logarithmic interval of k, by multiplying it with the phase volume factor. Thus we define

$$
\mathcal{P}_{\delta\phi_*}(k) \equiv \frac{4\pi k^3}{(2\pi)^3}P_{\delta\phi_*}(k) = \frac{H^2}{(2\pi)^2}\bigg|_{t_k}, \qquad (2.187)
$$

$$
\mathcal{P}_{\mathcal{R}_c}(k) \equiv \frac{4\pi k^3}{(2\pi)^3}P_{\mathcal{R}_c}(k) = \frac{1}{2\epsilon M_P^2}\frac{H^2}{(2\pi)^2}\bigg|_{t_k}. \qquad (2.188)
$$

We emphasize that all we need is to calculate $N(\phi)$ as a function of the background field ϕ and identify $\delta\phi$ in δN with the perturbation on a flat slice $\delta\phi_{\mathrm{fl}}$ when the scale of interest has just left the horizon, where the spectrum of $\delta\phi_{\mathrm{fl}}$ is given by (2.187).

The above results can be generalized for the case of multiple scalar fields. In the case of an $\mathcal{M}$-component scalar field ($I = 1, 2, \cdots, \mathcal{M}$), one obtains

$$
\mathcal{P}_{\mathcal{R}_c} = \sum_I N_{\phi_I}^2 \mathcal{P}_{\delta\phi_I} = \left(\frac{H}{2\pi}\right)^2\sum_I N_{\phi_I}^2\bigg|_{t_k}, \qquad (2.189)
$$

in which it is assumed that $\langle \delta\phi_I \delta\phi_J \rangle = (H^2/2k^3)\delta_{IJ}$, i.e. the $\delta\phi_I$ fluctuations are mutually uncorrelated.

In particular, having obtained the power spectrum as given in Eq. (2.188), we can calculate the spectral tilt of the power spectrum as

$$n_s - 1 \equiv \frac{d\,\ln\mathcal{P}_{\mathcal{R}_c}(k)}{d\,\ln k} = -2\epsilon - \eta = -6\epsilon + 2\eta_V. \qquad (2.190)$$

Due to a historical reason, the spectral index n_s is defined such that $n_s = 1$ corresponds to a scale-invariant spectrum. Assuming the single-field slow-roll inflation, the observational data requires $n_s \simeq 0.96$, i.e. a slightly red-tilted power spectrum for the comoving curvature perturbation [39].

2.4.2   Power spectrum and spectral index in δN formalism

Let us derive the power spectrum of the final, conserved comoving curvature perturbation when the adiabatic limit is reached. Here for the sake of clarity, we restore a general field space metric h_{IJ}, but we still assume that the field space is flat, i.e., the curvature is zero, $R^I_{JKL} = 0$, for simplicity.

Using Eq. (66), the power spectrum is expressed as

$$
\begin{aligned}
\frac{2\pi^2}{k^3}\mathcal{P}_{\mathcal{R}_c}\delta^3(\mathbf{k}-\mathbf{k}') &= \langle \mathcal{R}_{c\mathbf{k}}\mathcal{R}^*_{c\mathbf{k}'}\rangle(t_{\mathrm{fin}}) \\
&= \frac{\partial N}{\partial\phi^I}\frac{\partial N}{\partial\phi^J}\langle \delta\phi^I_{f\mathbf{k}}\delta\phi^{J*}_{f\mathbf{k}'}\rangle(t_{\mathrm{ini}}).
\end{aligned}
\qquad (2.191)
$$

Assuming the slow-roll conditions (2.74), the field fluctuations $\delta\phi^I_{f\mathbf{k}}$ can be evaluated in the standard manner.

Taking the initial time $t = t_{\mathrm{ini}}$ ($\eta = \eta_{\mathrm{ini}}$) to be the time when the scales of interest are outside the Hubble horizon, we have

$$\delta\phi^I_{f\mathbf{k}} = \phi^{I\alpha}_k a^\alpha_{\mathbf{k}}, \qquad (2.192)$$

where $a^\alpha_{\mathbf{k}}$ is an ortho-normalized, complex random Gaussian variable,

$$\langle a^\alpha_{\mathbf{k}} a^{\beta*}_{\mathbf{k}'}\rangle = \delta^{\alpha\beta}\delta^3(\mathbf{k}-\mathbf{k}') \qquad (2.193)$$

and α runs over the number of scalar field components, that is $\alpha = 1, 2, 3, \cdots, \mathcal{M}$. It is noteworthy that, in the slow-roll approximation, the decaying mode becomes negligible outside the horizon; however for general cases this is not the case and N is a function of both initial amplitude of the fields and their derivatives, say $N = N\left(\phi^I, \dot{\phi}^I\right)$ and Eq. (2.191) should be modified accordingly. We address this issue in more details in Chapter 5. Nevertheless, here we focus on the models subjected to slow-roll conditions. $\phi_k^{I\alpha}$ is the amplitude of each independent fluctuation which we take to be real, and which satisfies

$$\sum_\alpha \phi_k^{I\alpha} \phi_k^{J\alpha} = \frac{H^2}{2k^3} \left(h^{IJ} + \epsilon^{IJ}\right), \tag{2.194}$$

where ϵ^{IJ} is a quantity of the order of the slow-roll parameters which is slowly varying in time. From the above, we have

$$\langle \delta\phi_{f\mathbf{k}}^I \delta\phi_{f\mathbf{k}'}^{J*} \rangle = \sum_\alpha \phi_k^{I\alpha} \phi_k^{J\alpha} \delta^3 \left(\mathbf{k} - \mathbf{k}'\right)$$
$$= \frac{H^2}{2k^3} \left(h^{IJ} + \epsilon^{IJ}\right) \delta^3 \left(\mathbf{k} - \mathbf{k}'\right) + \mathcal{O}\left(\epsilon, \eta\right) \tag{2.195}$$

Plugging the above into Eq. (212), we obtain

$$\mathcal{P}_{\mathcal{R}_c}(k) = \frac{k^3}{2\pi^2} \frac{\partial N}{\partial \phi^I} \frac{\partial N}{\partial \phi^J} \sum_\alpha \phi_k^{I\alpha} \phi_k^{J\alpha} \bigg|_{t=t_{\text{ini}}}$$
$$= \frac{\partial N}{\partial \phi^I} \frac{\partial N}{\partial \phi^J} \frac{H^2}{4\pi^2} \left(h^{IJ} + \epsilon^{IJ}\right) \bigg|_{t=t_{\text{ini}}}. \tag{2.196}$$

As is clear from this formula, the spectrum is scale-invariant at leading order in the slow-roll parameters. The k-dependence is contained in the small correction term ϵ^{IJ}.

Now we turn to the evaluation of the spectral index of the curvature perturbation, which is expected to be of the order of the slow-roll parameters. It is defined as

$$n_s - 1 = \frac{d\ln \mathcal{P}_{\mathcal{R}_c}}{d\ln k}. \tag{2.197}$$

From the first line of Eq. (2.196), we have

$$n_s - 1 = 3 + \frac{k^3}{2\pi^2 \mathcal{P}_{\mathcal{R}_c}} \frac{\partial N}{\partial \phi^I} \frac{\partial N}{\partial \phi^J} \frac{\partial}{\partial \ln k} \sum_\alpha \phi_k^{I\alpha} \phi_k^{J\alpha} \bigg|_{N=\text{const.}} . \qquad (2.198)$$

To evaluate the k-derivative, we make use of the fact that ϵ^{IJ} is small and slowly varying in time for fixed k/a. Recalling $N \propto \ln a$, the k-derivative may be rewritten as

$$\frac{\partial}{\partial \ln k}\bigg|_{a=\text{const.}} = \frac{\partial}{\partial N}\bigg|_{k/a=\text{const.}} - \frac{\partial}{\partial N}\bigg|_{k=\text{const.}} . \qquad (2.199)$$

Then we can separate the k-derivative of $\sum_\alpha \phi_k^{I\alpha} \phi_k^{J\alpha}$ into the part due to the variation in the initial amplitude with the aid of Eq. (2.76),

$$\frac{\partial}{\partial N}\bigg|_{k/a=\text{const.}} \left(\sum_\alpha \phi_k^{I\alpha} \phi_k^{J\alpha} \right) = \left(2\frac{\dot{H}}{H^2} - 3 \right) \sum_\alpha \phi_k^{I\alpha} \phi_k^{J\alpha}, \quad (2.200)$$

where one can neglect the derivative of ϵ^{IJ}, and the part due to the time evolution of $\phi_k^{I\alpha}$,

$$\frac{\partial \phi_k^{I\alpha}}{\partial N}\bigg|_{k=\text{const.}} = \left(\delta_{JK} \phi_N^I \phi_N^K - \frac{V^I{}_J}{V} \right) \phi_k^{J\alpha}. \qquad (2.201)$$

Then, the spectral index is evaluated as

$$\begin{aligned}
n_s - 1 &= 3 + \frac{k^3}{2\pi^2 \mathcal{P}_{\mathcal{R}_c}} \frac{\partial N}{\partial \phi^I} \frac{\partial N}{\partial \phi^J} \left(2\frac{\dot{H}}{H^2} - 3 \right) \sum_\alpha \phi_k^{I\alpha} \phi_k^{J\alpha} \\
&\quad -2 \frac{k^3}{2\pi^2 \mathcal{P}_{\mathcal{R}_c}} \frac{\partial N}{\partial \phi^I} \frac{\partial N}{\partial \phi^J} \left(h_{KL} \phi_N^I \phi_N^L - \frac{V^I{}_K}{V} \right) \sum_\alpha \phi_k^{K\alpha} \phi_k^{J\alpha} \\
&= 2\frac{\dot{H}}{H^2} - 2 \frac{\left(\phi_N^I \phi_N^J - \dfrac{V^{IJ}}{V} \right) \dfrac{\partial N}{\partial \phi^I} \dfrac{\partial N}{\partial \phi^J}}{h^{KL} \dfrac{\partial N}{\partial \phi^K} \dfrac{\partial N}{\partial \phi^L}} .
\end{aligned}$$

$$(2.202)$$

The above formula can be further simplified under the slow-roll approximation. Using the first of Eq. (2.74) and the slow-roll equations of motion (2.75) we express $\dot{H}/H^2$ and ϕ_N^I in terms of derivatives of the potential to find

$$n_s - 1 = \frac{\left[2(\ln V)^{IJ} - h^{IJ} h_{KL} (\ln V)^K (\ln V)^L\right] \dfrac{\partial N}{\partial \phi^I} \dfrac{\partial N}{\partial \phi^J}}{h^{KL} \dfrac{\partial N}{\partial \phi^K} \dfrac{\partial N}{\partial \phi^L}}. \qquad (2.203)$$

Before closing this subsection, we also mention the results for a curved field space metric. In this case, it is known that the curvature term appears in the expression of the power spectrum. Namely, the spectrum index is given by [16]

$$n_s - 1 =$$

$$\frac{\left[2(\ln V)^{IJ} + \left(\dfrac{2}{3} h^{JM} R^I_{KML} - h^{IJ} h_{KL}\right)(\ln V)^K (\ln V)^L\right] \dfrac{\partial N}{\partial \phi^I} \dfrac{\partial N}{\partial \phi^J}}{\delta^{KL} \dfrac{\partial N}{\partial \phi^K} \dfrac{\partial N}{\partial \phi^L}},$$

$$(2.204)$$

where the curvature and Christoffel symbols of the scalar field metric are defined as

$$R^I_{JKL} \equiv \Gamma^I_{JL,K} - \Gamma^I_{JK,L} + \Gamma^I_{KM} \Gamma^M_{JL} - \Gamma^I_{LM} \Gamma^M_{JK}, \qquad (2.205)$$

$$\Gamma^I_{JK} \equiv \frac{1}{2} h^{IL} \left(h_{LJ,K} + h_{LK,J} - h_{JK,L}\right). \qquad (2.206)$$

2.4.3 *Non-Gaussianities*

Cosmological observations indicate that the primordial fluctuations are highly Gaussian [40,41]. Namely the main statistical information is captured by the two-point correlation functions while the three-point and higher order (1-particle irreducible, or intrinsic) correlations are nearly zero (to be quantified below). Any deviation from Gaussianity is called "non-Gaussianity" [42,43]. In a non-Gaussian distribution, higher order correlation functions (to be more precise the connected ones) have non-zero values. In the momentum space, the higher order correlations are called the bispectrum, trispectrum and so on. But often one refers to the bispectrum as non-Gaussianity as it usually makes the dominant contribution. Intuitively speaking, non-Gaussianity is a measure of deviation from linearity or, alternatively, a measure of interaction in the system. Therefore, any detection of non-Gaussianity or otherwise can put strong constraints on the dynamics of inflation and the subsequent evolution of curvature perturbations after horizon crossing. In this view non-Gaussianity is a powerful tool for discriminating various inflationary models.

Historically, non-Gaussianity is parametrized by the non-linear parameter f_{NL} as follows

$$\Phi\left(\mathbf{x}\right) = \Phi_g\left(\mathbf{x}\right) + f_{NL}\left(\Phi_g^2\left(\mathbf{x}\right) - \langle\Phi_g\left(\mathbf{x}\right)\rangle^2\right) + \cdots, \qquad (2.207)$$

where Φ is the curvature perturbation on the Newton (shear-free) slices (called the Bardeen potential) and Φ_g represents its Gaussian (linear) part. In the matter-dominated stage, which is a good approximation at the time of photon–baryon decoupling, Φ is related to the conserved comoving curvature perturbation $\mathcal{R}_c$ as $\Phi = \frac{3}{5}\mathcal{R}_c$. Therefore the above in terms of $\mathcal{R}_c$ becomes

$$\mathcal{R}_c\left(\mathbf{x}\right) = \mathcal{R}_{c,g}\left(\mathbf{x}\right) + \frac{3}{5}f_{NL}\left(\mathcal{R}_{c,g}^2\left(\mathbf{x}\right) - \langle\mathcal{R}_{c,g}\left(\mathbf{x}\right)\rangle^2\right) + \cdots. \qquad (2.208)$$

This form is often used to describe the effect of non-Gaussianity, and in fact is the form that can be most conveniently discussed in terms of the δN formalism.

However, we note that the general shape and amplitude of non-Gaussianity is not entirely captured by the single parameter f_{NL}. Specifically, the ansatz (2.207) or (2.208) is valid for the so-called local-type non-Gaussianity. There are other types of non-Gaussianity which may not be captured by this simple ansatz. Nevertheless, the form (2.207) is still useful in making a crude, order-of-magnitude estimate of the level of non-Gaussianity. Comparing the non-Gaussian term $f_{NL}\Phi_g^2(\mathbf{x})$ with the Gaussian term $\Phi_g(\mathbf{x})$, the level of non-Gaussianity, that is, the ratio of the non-linear term to the linear term, may be estimated as $f_{NL}\Phi_g(\mathbf{x})$. From the COBE normalization we have $\Phi_g(\mathbf{x}) \sim 10^{-5}$ (of the order of the primordial fluctuations on the CMB), this implies $f_{NL}\Phi_g(\mathbf{x}) \sim 10^{-5}f_{NL}$. Therefore, the level of non-Gaussianity would be very small unless $f_{NL} \gtrsim 10^4$. This suggests that the CMB fluctuations are indeed highly Gaussian in general. In fact, single-field slow-roll models of inflation generically predict extremely small non-Gaussianity, $f_{NL} \sim \mathcal{O}(\epsilon, \eta)$ [44]. Therefore, any detection of non-Gaussianity even at the order of unity, $f_{NL} \sim 1$, will rule out *all* single-field *slow-roll* models of inflation.

The general form of non-Gaussianity to first non-trivial order is

$$
\mathcal{R}_c(\mathbf{x}) = \mathcal{R}_{c,g}(\mathbf{x}) \\
+ \int d^3y \, d^3z \, K(\mathbf{x}; \mathbf{y}, \mathbf{z}) \left(\mathcal{R}_{c,g}(\mathbf{y}) \mathcal{R}_{c,g}(\mathbf{z}) - \langle \mathcal{R}_{c,g}(\mathbf{y}) \mathcal{R}_{c,g}(\mathbf{z}) \rangle \right),
$$

$$(2.209)$$

where $\mathcal{R}_{c,g}$ is the leading order Gaussian part of the curvature perturbation. The simple ansatz (2.207) corresponds to the case,

$$
K(\mathbf{x}; \mathbf{y}, \mathbf{z}) = \frac{3}{5} f_{NL} \delta^3(\mathbf{y} - \mathbf{x}) \delta^3(\mathbf{z} - \mathbf{x}).
$$

$$(2.210)$$

Instead of directly dealing with the above form, which actually contains stochastic variables, it is often more convenient to characterize the non-Gaussianity by the n-point correlation functions, which are just numbers. A Gaussian distribution is characterized solely by the two-point correlation function, $\langle \mathcal{R}_c(\mathbf{x}) \mathcal{R}_c(\mathbf{x}') \rangle$, while a non-Gaussian distribution will have non-trivial "connected" n-point correlation functions.

Although it is perfectly fine to consider the real space n-point functions, when we test the predictions of inflationary models against observational data, it is more convenient to characterize the non-Gaussianity in terms of the momentum space n-point functions. In this case, the n-point function is generally called the m-spectra with $m = 1, 2, \cdots, n - 1$, namely, the spectrum ($n = 2$), bispectrum ($n = 3$), trispectrum ($n = 4$), etc.

To be more specific, let us consider some of the lower order correlation functions in momentum space. We focus on those of the comoving curvature perturbation. The 2-point function is given in terms of the spectrum $P_{\mathcal{R}_c}(\mathbf{k})$ as

$$
\langle \mathcal{R}_c(\mathbf{k}) \mathcal{R}_c(\mathbf{k}') \rangle = (2\pi)^3 \delta^3(\mathbf{k} + \mathbf{k}') P_{\mathcal{R}_c}(k),
$$

$$(2.211)$$

where we have assumed the spatially homogeneous and isotropic statistical distribution of $\mathcal{R}_c(\mathbf{x})$, and as before the Fourier components are defined as

$$\mathcal{R}_c(\mathbf{k}) \equiv \int d^3x \, \mathcal{R}_c(\mathbf{x}) \, e^{i\mathbf{k}\cdot\mathbf{x}}. \qquad (2.212)$$

The delta function represents the conservation of momentum which is a manifestation of the translational invariance. The next order spectrum, the bispectrum is defined as

$$\langle \mathcal{R}_c(\mathbf{k}_1) \mathcal{R}_c(\mathbf{k}_2) \mathcal{R}_c(\mathbf{k}_3) \rangle = (2\pi)^3 \delta^3(\mathbf{k}_1 + \mathbf{k}_2 + \mathbf{k}_3) \, B(\mathbf{k}_1, \mathbf{k}_2, \mathbf{k}_3), \qquad (2.213)$$

where

$$B(\mathbf{k}_1, \mathbf{k}_2, \mathbf{k}_3) = \frac{6}{5} f_{NL}(\mathbf{k}_1, \mathbf{k}_2) \, P_{\mathcal{R}_c}(k_1) P_{\mathcal{R}_c}(k_2) + 2\,\mathrm{perm}. \qquad (2.214)$$

Again the delta function is due to the translational invariance. Computing the 3-point function for (2.207), one can easily sees that the case of a local-type non-Gaussianity corresponds to a constant f_{NL}. In this sense, the above expression can be regarded as the generalized f_{NL} in momentum space.

To further characterize the bispectrum, it is convenient to introduce the notion of the shape function by focusing on the shape of the triangle formed by $\mathbf{k}_1$, $\mathbf{k}_2$ and $\mathbf{k}_3$, modulo its amplitude. For convenience, let us consider $\mathbf{k}_2/k_1$ and $\mathbf{k}_3/k_1$, where $k_1 = |\mathbf{k}_2 + \mathbf{k}_3|$. The number of the independent degrees of freedom is 5. Furthermore, spatial homogeneity and isotropy imply the rotational invariance, which reduces 3 more degrees of freedom. Thus we end up with 2 degrees of freedom, which are usually represented by the two ratios k_2/k_1 and k_3/k_1 [45].

The shape function is a key parameter for discriminating among different models of inflation. Different shapes are closely related to different non-linearities that can generate such non-Gaussian fingerprints on CMB maps. It is known that the general shape of non-Gaussianities can be represented by three shapes called the local, equilateral and orthogonal shapes. The first shape has its peak for squeezed triangles $k_1 \ll k_2 \simeq k_3$ while the equilateral shape is peaked for equal momenta,

$k_1 = k_2 = k_3$. The orthogonal shape is orthogonal to the equilateral shape in the sense of a particular 3D scalar product introduced in Ref. [46]. Below, however, we focus on the local-type non-Gaussianity, since it is the type of non-Gaussianity that can be captured by the δN formalism. The other types which correspond to f_{NL} with non-trivial dependence on the momenta can be generated only on subhorizon scales where couplings between different momenta may appear through causal processes.

2.4.4 *Implementation of δN formalism for non-Gaussianity*

The δN formula, $\mathcal{R}_c = \delta N$, is valid to all orders in perturbation, by appropriately defining the non-linear curvature perturbation as discussed before. Therefore, one can expand $N(\phi)$ to higher orders in the field perturbation to calculate higher order correlation functions of the curvature perturbation.

The bispectrum of the primordial curvature perturbations can be obtained by expanding $N(\phi)$ to second order in field perturbation,

$$
\begin{aligned}
\mathcal{R}_c(\mathbf{x}) = \delta N(\mathbf{x}) &= N(\phi_* + \delta\phi_*(\mathbf{x})) - N(\phi_*) \\
&= N_I \delta\phi_*^I(\mathbf{x}) + \frac{1}{2} N_{IJ} \delta\phi_*^I(\mathbf{x}) \delta\phi_*^J(\mathbf{x}) + \mathcal{O}(\delta\phi^3),
\end{aligned}
\tag{2.215}
$$

where $N_I \equiv \partial N/\partial \phi^I$, $N_{IJ} \equiv \partial^2 N/\partial \phi^I \partial \phi^J$, and as before $\delta\phi_*^I$ is the vacuum fluctuation evaluated on flat slicing and the right-hand side is to be evaluated during inflation, at an epoch $t = t_*$ when the scale of interest has become larger than the Hubble horizon scale. In terms of the comoving wavenumber, this implies that we focus on the modes $k < k_*$ where k_* is the wavenumber that crosses the horizon at $t = t_*$, $k_* = H(t_*) = a(t_*)H(t_*)$.

The above discussion was given in the real space. However, to compare the predicted non-Gaussianities with observations, it is more convenient to consider the Fourier space n-point functions as discussed in the previous section. For a slow-rolling scalar field, the time variation of $\delta\phi_{\mathrm{fl}}$ is negligible over a few e-folds, $|\Delta\delta\phi_{\mathrm{fl}}| = |d\delta\phi_{\mathrm{fl}}/dN|\Delta N \approx |(\eta/2)\Delta N \delta\phi_{\mathrm{fl}}| \ll |\delta\phi_{\mathrm{fl}}|$ for, say, $\Delta N \lesssim 10$. Therefore at leading order, we may ignore the evolution of $\delta\phi_{\mathrm{fl}}$ on superhorizon scales. In other words, we may take the Fourier transform of Eq. (2.215) to obtain

$$\mathcal{R}_c(\mathbf{k}) = N_I \delta\phi_*(\mathbf{k}) + \frac{1}{2} N_{IJ} \int \frac{d^3 p}{(2\pi)^3} \delta\phi_*^I(\mathbf{k} - \mathbf{p}) \delta\phi_*^J(\mathbf{p}) + \mathcal{O}(\delta\phi^3),$$

$$(2.216)$$

where it is understood that the right-hand side is evaluated at $N = N(t_*)$ and the momentum integral extends only over superhorizon modes.

Taking the 3-point function of the above, the bispectrum of the curvature perturbation is computed to be

$$\langle \mathcal{R}_c(\mathbf{p}_1) \mathcal{R}_c(\mathbf{p}_2) \mathcal{R}_c(\mathbf{p}_3) \rangle = (2\pi)^3 \delta(\mathbf{p}_1 + \mathbf{p}_2 + \mathbf{p}_3)$$

$$\times \left[N_I N_J N_K B_{\delta\phi_*}(\mathbf{p}_1, \mathbf{p}_2, \mathbf{p}_3) + N_I N_J N_{KL} h^{IK} h^{JL} \right.$$

$$\left. \left(P_{\delta\phi_*}(p_1) P_{\delta\phi_*}(p_2) + P_{\delta\phi_*}(p_2) P_{\delta\phi_*}(p_3) + P_{\delta\phi_*}(p_3) P_{\delta\phi_*}(p_1) \right) \right],$$

$$(2.217)$$

where the 2-point and 3-point functions of $\delta\phi_*$ are assumed to be given as

$$\langle \delta\phi_*^I(\mathbf{p}_1) \delta\phi_*^J(\mathbf{p}_2) \rangle = (2\pi)^3 \delta(\mathbf{p}_1 + \mathbf{p}_2) h^{IJ} P_{\delta\phi_*}(p_1),$$

$$\langle \delta\phi_*^I(\mathbf{p}_1) \delta\phi_*^J(\mathbf{p}_2) \delta\phi_*^K(\mathbf{p}_3) \rangle = (2\pi)^3 \delta(\mathbf{p}_1 + \mathbf{p}_2 + \mathbf{p}_3) B_{\delta\phi_*}^{IJK}(p_1, p_2, p_3),$$

$$(2.218)$$

and all the terms on the right-hand side are to be evaluated at the horizon crossing time of the largest wavenumber, say at $a(t_{p_1}) H(t_{p_1}) = p_1$ for $p_1 \geq p_2 \geq p_3$.

In the squeezed limit, $p_1 \approx p_2 \gg p_3$, Eq. (2.217) reduces to

$$\langle \mathcal{R}_c(\mathbf{p}_1) \mathcal{R}_c(\mathbf{p}_2) \mathcal{R}_c(\mathbf{p}_3) \rangle_{p_1 \approx p_2 \gg p_3} \approx (2\pi)^3 \delta(\mathbf{p}_1 + \mathbf{p}_2 + \mathbf{p}_3)$$

$$\times \left[N_I N_J N_K B^{IJK}(p_1, p_1, p_3) + 2 N_I N_J N_{KL} h^{IK} h^{JL} P_{\delta\phi_*}(p_1) P_{\delta\phi_*}(p_3) \right],$$

$$(2.219)$$

where we have assumed that the spectrum is almost scale-invariant, $P_{\delta\phi_*}(k) \sim k^{-3}$. Comparing the above with Eqs. (2.213) and (2.214), and noting the relation $\mathcal{R}_c = N_I \delta\phi_*^I$ at leading order, we find

$$\frac{6}{5} f_{NL}(p) \underset{p \equiv p_1 \approx p_2 \gg p_3}{=} \left[\frac{N_I N_J N_{KL} h^{IK} h^{JL}}{(N_I N_J h^{IJ})^2} + \frac{N_I N_J N_K B^{IJK}}{2(N_I N_J h^{IJ})^2 P_{\delta\phi_*}(p_3) P_{\delta\phi_*}(p)} \right]. \tag{2.220}$$

The first term on the right-hand side is the contribution from the δN formalism, while the second term is that from the intrinsic non-Gaussianity of $\delta\phi_*$. Note that rigorously speaking the δN contribution here is different from f_{NL} given by Eq. (2.221) in that it is now defined in the Fourier space.

For single-field slow-roll inflation, comparing Eq. (2.215) with Eq. (2.208), we find that f_{NL} is given by

$$\frac{3}{5} f_{NL}^{\delta N} = \frac{N_{,\phi\phi}}{2N_{,\phi}^2} = \frac{\eta}{4}, \tag{2.221}$$

provided that $\delta\phi_*$ is Gaussian. We have attached the superscript δN to indicate that it is the part which can be computed by the δN formalism. In reality there is a small intrinsic non-Gaussianity in $\delta\phi_*$. The in-in formalism computation gives [47]

$$B\left(p_1, p_2, p_3\right)_{p_3 \ll p_1, p_2} = 2\epsilon N_{,\phi} P_{\delta\phi_*}(p_3) P_{\delta\phi_*}(p_1). \tag{2.222}$$

Plugging this into Eq. (2.220), the total f_{NL} in the squeezed limit $p_3 \ll p$ is given by the sum of the δN part and the intrinsic part,

$$\frac{6}{5} f_{NL}(p) = \frac{6}{5} f_{NL}^{\delta N}(p) + \frac{6}{5} f_{NL}^{\text{int}}(p) = \frac{\eta}{2} + \epsilon, \tag{2.223}$$

where, as noted below Eq. (2.217), the right-hand side is to be evaluated at the horizon crossing time $t = t_p$ of p; $a\left(t_p\right) H\left(t_p\right) = p$.

Comparing the above with Eq. (2.190), we find that f_{NL} in the squeezed limit is related to the spectral index of the comoving curvature perturbation as [44]

$$\frac{12}{5} f_{NL} = 1 - n_s = -\frac{d\ln \mathcal{P}_{\mathcal{R}_c}(k)}{d\ln k}. \tag{2.224}$$

It has been shown that this relation holds very generically for any type of single-field slow-roll models, not necessarily of canonical type, though f_{NL} itself will acquire additional contributions due to the existence of more parametric degrees of freedom in non-canonical models [48].

Before closing this subsection, it may be worth noting that the above decomposition of f_{NL} into the intrinsic and δN parts depends actually on the definition of the scalar field. For example, if one introduces a non-canonical scalar field χ by $\phi = F(\chi)$, we have

$$\delta\phi = \frac{dF}{d\chi}\delta\chi + \frac{1}{2}\frac{d^2 F}{d\chi^2}\delta\chi^2 + \cdots. \tag{2.225}$$

This gives

$$\begin{aligned}
\delta N(\chi) &= \frac{\partial N}{\partial\chi}\delta\chi + \frac{1}{2}\frac{\partial^2 N}{\partial\chi^2}\delta\chi^2 + \cdots \\
&= \frac{\partial N}{\partial\phi}\delta\phi + \frac{1}{2}\left(\frac{\partial^2 N}{\partial\phi^2} + \frac{F''}{F'^2}\frac{\partial N}{\partial\phi}\right)\delta\phi^2 + \cdots.
\end{aligned} \tag{2.226}$$

Hence $f_{NL}^{\delta N}$ will have an additional term,

$$\frac{6}{5}\Delta f_{NL} = \frac{F''}{F'^2 N_{,\phi}}. \tag{2.227}$$

This implies that one may actually choose a new definition of the scalar field such that it is purely Gaussian in the squeezed limit [49]. Here in order not to get confused, we note that the total f_{NL} which is the sum of f_{NL}^{int} and $f_{NL}^{\delta N}$ is invariant under this redefinition of the field. It is the f_{NL}^{int} part (or the $f_{NL}^{\delta N}$ part) which can be set to zero by redefinition of the field.

CHAPTER **3**

Application of δN formalism: Warm-up studies

In Chapter 2 we have presented the general formulation of the δN approach to cosmological perturbations. In this chapter as a warm-up, we present simple applications of the δN formalism.

This chapter has two separate examples. In the first example, we show how the δN formalism can be used to calculate the power spectrum of the curvature perturbation in single-field slow-roll inflation. In the second example, we review the curvaton scenario as a novel mechanism for generating the curvature perturbation and non-Gaussianities, and use the δN formalism to calculate the power spectrum and bispectrum.

3.1 A specific model: Chaotic inflation

Having presented the analysis for general single-field models, here we apply the δN formalism to the simplest class of models, the chaotic potential, as an illustration:

$$V(\phi) = \frac{m^2\phi^2}{2n}\left(\frac{\phi^2}{M_P^2}\right)^{n-1} \quad (n = 1, 2, \cdots).$$

(3.1)

The slow-roll parameters for this class of models are

$$\epsilon_V = \frac{M_P^2}{2}\left(\frac{V_{,\phi}}{V}\right)^2 = 2n^2\frac{M_P^2}{\phi^2}, \quad \eta_V = M_P^2\frac{V_{,\phi\phi}}{V} = 2n\left(2n-1\right)\frac{M_P^2}{\phi^2}.$$

(3.2)

Consequently, the slow-roll conditions require $\phi \gg M_P$, i.e. the field value is super-Planckian. In the literature these types of models are called large-field models. Also, in this class of models one easily sees that inflation ends at $\phi = \mathcal{O}(M_P)$ unless n is extremely large.

The slow-roll equations of motion, Eqs.(2.24) and (2.25) become

$$3M_P^2H^2 = \frac{m^2\phi^2}{2n}\left(\frac{\phi^2}{M_P^2}\right)^{n-1}, \quad \dot{\phi} = -\frac{m^2\phi}{3H}\left(\frac{\phi^2}{M_P^2}\right)^{n-1}.$$

(3.3)

What we are interested in is the number of e-folds $N(\phi)$ from the time $t = t(\phi)$ until the end of inflation, given by Eq. (2.30). In the present case it is

$$N = \frac{1}{M_P^2}\int_{\phi_f}^{\phi}\frac{V}{V_{,\phi}}d\phi = \frac{1}{4n}\frac{\phi^2 - \phi_f^2}{M_P^2} \simeq \frac{1}{4n}\frac{\phi^2}{M_P^2},$$

(3.4)

where the last approximate equality follows from the fact that $\phi^2 \gg M_P^2$ while $\phi_f^2 = \mathcal{O}(M_P^2)$. In passing, we note that for the quadratic potential model, $n = 1$, $\eta_V = \epsilon_V = 2M_P^2/\phi^2$. Hence $\phi = \phi_f = \sqrt{2}M_P$ may be identified as the end of inflation.

Now, following the general prescription of the δN formalism, the linear variation of Eq. (3.4) can be used to find the linear curvature perturbation,

$$\mathcal{R}_c = \delta N = \frac{1}{2n}\frac{\phi}{M_P^2}\delta\phi_*.$$

(3.5)

Consequently, the curvature perturbation power spectrum calculated from Eq. (2.188) is obtained to be

$$\mathcal{P}_{\mathcal{R}_c}(k) = \frac{\phi^2}{(2n)^2 M_P^2}\frac{H^2}{4\pi^2 M_P^2}\bigg|_{t_k} = \frac{m^2}{96n^3\pi^2 M_P^2}\left(\frac{\phi^2}{M_P^2}\right)^{n+1}\bigg|_{t_k}.$$

(3.6)

In terms of the number of e-folds N, this may be re-expressed as

$$\mathcal{P}_{\mathcal{R}_c}(k) = \frac{2(4n)^{n-2}}{3\pi^2}\frac{m^2}{M_P^2}(N_k)^{n+1}, \tag{3.7}$$

where $N_k = N(\phi(t_k))$. If we were to fit this with the observed amplitude, $\mathcal{P}_{\mathcal{R}_c} \simeq 2 \times 10^{-9}$ (at $k_* = 0.05$ Mpc^{-1}), the inflaton mass must be $m \simeq 6 \times 10^{-6}$ for the quadratic potential $n = 1$ if $N_k \simeq 60$.

3.2 Curvaton model

Usually it is assumed that the superhorizon curvature perturbations are generated during inflation by the inflaton field. This is perhaps most natural and economical. However, this is not the only possibility. Indeed, one can consider the situations in which the superhorizon curvature perturbations are generated by fields other than the inflaton. However, the crucial requirement is that the initial seeds of these large scale perturbations should be generated *during* inflation. Otherwise, there would be no casual mechanism to generate these perturbations on superhorizon scales.

The curvaton scenario [50–53] is one of this kind. In this scenario it is assumed that there are two fields during inflation, the inflaton ϕ and another scalar field σ, dubbed curvaton, which is assumed to be very light during inflation. Moreover, it is assumed that the inflaton potential drives the background inflation. However, the amplitude of the curvature perturbation generated by the inflaton is assumed to be too small to explain the observed value. For example, one may consider a chaotic inflation model with the inflaton mass many orders of magnitude smaller than the required mass $m_\phi \sim 10^{-6} M_P$ for the observed amplitude of the curvature perturbation.

As the curvaton field does not take part in the background dynamics it does not have any contribution to the curvature perturbation during inflation. In other words, it is an isocurvature field. In this scenario, after inflation the inflaton field either starts to oscillate and decay to radiation or directly decays to radiation, while the curvaton still remains at its value until the Hubble parameter decreases to become comparable to the curvaton mass, $H \sim m_\sigma$. After this epoch, the universe consists of the radiation from the inflaton decay, $\rho_\gamma \propto a^{-4}$, and the oscillating curvaton that behaves like dust, $\rho_\sigma \propto a^{-3}$.

The curvaton is assumed to decay eventually to radiation. Let us denote this decay rate by Γ. The consistency of this picture requires $\Gamma < m_\sigma$. Once the expansion rate drops below Γ, the curvaton decays to radiation. For simplicity, we employ the sudden decay approximation. Namely, we assume the curvaton decays to radiation instantaneously when $H = \Gamma$. If the curvaton is the dominant component of the universe at the time of decay, the resulting curvature perturbation will be dominated by that due to the vacuum fluctuations of the curvaton.

With this introduction, we are ready to apply the δN formalism to the curvaton scenario. For this purpose, let us slightly generalize the δN formalism to the case of a multi-component fluid/field. First of all, using the fact that the comoving and uniform density slicings are identical to each other at leading order in gradient expansion, we replace the final comoving slice by the uniform density slice [18]. As discussed in Chapter 2, the δN formula says the curvature perturbation on the comoving slices, which is supposedly conserved on superhorizon scales in the adiabatic limit, is equal to the perturbation in the number of e-folds between the final comoving slice and the initial flat slice,

$$\delta N (t_1, t_2; \mathbf{x}) = \mathcal{R}_c (t_2, \mathbf{x}) . \tag{3.8}$$

Even if we have a multi-component fluid/field, if different components do not interact with each other, other than gravitationally, and each component is in the adiabatic limit, one may then introduce the δN formula for each component by replacing the comoving slice with uniform density slice and by using the energy conservation law of each component [18]. This means that even at a stage before the universe has reached the adiabatic limit, we may have a way to compute the curvature perturbation on a given slice by using only the background equations of motion, provided that all the components have reached the adiabatic limit.

Let us consider the energy conservation law of a fluid/field labeled by A at leading order in gradient expansion,

$$\frac{d\rho^A}{d\tau} + K\left(\rho^A + P^A\right) = 0, \tag{3.9}$$

where $K = d\ln\sqrt{\gamma}/d\tau$ with $\sqrt{\gamma} = a^3\left(t\right)\exp\left(3\mathcal{R}\left(\mathbf{x},t\right)\right)$ being the 3-volume element and τ being the proper time in the fluid/field rest frame, $d\tau = \alpha dt$. For convenience, we take the geometry at $\mathbf{x} = 0$ as fiducial and set $\mathcal{R}(0,t) = 0$. Rewriting this for K and integrating it along the proper time in the rest frame, we find

$$\frac{1}{3}\int_{\tau_1}^{\tau_2} d\tau K\left(\mathbf{x},\tau\right) = \ln\frac{a\left(t_2\right)}{a\left(t_1\right)} + \mathcal{R}\left(\mathbf{x},t_2\right) - \mathcal{R}\left(\mathbf{x},t_1\right)$$

$$= -\int_{t_1}^{t_2} dt\frac{\dot{\rho}^A}{3\left(\rho^A + P^A\right)}, \tag{3.10}$$

where $\dot{\rho} = d\rho/dt$.

Let us further assume that each component is either an adiabatic fluid where $P^A = P\left(\rho^A\right)$ or a field in the adiabatic limit where ρ^A and P^A are determined by the field value, $\rho^A = \rho\left(\phi^A\right)$ and $P^A = P\left(\phi^A\right)$. In this case the right-hand side can be integrated,

$$\int_{t_1}^{t_2} dt\frac{\dot{\rho}^A}{3\left(\rho^A + P^A\right)} = \frac{1}{3}\int_{\rho(\mathbf{x},t_1)}^{\rho(\mathbf{x},t_2)}\frac{d\rho}{\rho + P}, \tag{3.11}$$

where here and below we omit the label A for notational simplicity unless confusion may arise. Then subtracting the contribution of the fiducial part we obtain from Eq. (3.10),

$$\mathcal{R}\left(\mathbf{x},t_2\right) - \mathcal{R}\left(\mathbf{x},t_1\right) = -\frac{1}{3}\int_{\rho(0,t_2)}^{\rho(\mathbf{x},t_2)}\frac{d\rho}{\rho + P} + \frac{1}{3}\int_{\rho(0,t_1)}^{\rho(\mathbf{x},t_1)}\frac{d\rho}{\rho + P}, \tag{3.12}$$

or

$$\mathcal{R}\left(\mathbf{x},t_2\right) + \frac{1}{3}\int_{\rho(0,t_2)}^{\rho(\mathbf{x},t_2)}\frac{d\rho}{\rho + P} = \mathcal{R}\left(\mathbf{x},t_1\right) + \frac{1}{3}\int_{\rho(0,t_1)}^{\rho(\mathbf{x},t_1)}\frac{d\rho}{\rho + P}. \tag{3.13}$$

As clearly seen, this means the quantity,

$$\zeta^A \equiv \mathcal{R}(\mathbf{x},t) + \frac{1}{3} \int_{\rho^A(0,t)}^{\rho^A(\mathbf{x},t)} \frac{d\rho^A}{\rho^A + P^A}, \tag{3.14}$$

is conserved, where we have reintroduced the label A to clarify that this applies to each component. Furthermore, the fact that it is conserved in time implies it is invariant under a change of time-slicing. In particular, if we take the slicing on which ρ^A is homogeneous, we immediately find

$$\mathcal{R}_{\rho^A}^A(\mathbf{x},t) = \zeta^A(\mathbf{x}), \tag{3.15}$$

where $\mathcal{R}_{\rho^A}^A$ denotes the curvature perturbation on the uniform ρ^A slice. This is the non-linear generalization of the conserved curvature perturbation on uniform density slices. Then the non-linear δN formula for each component is simply given by

$$\delta N^A(t_i, t_f; \mathbf{x}) = \mathcal{R}_{\rho^A}^A(\mathbf{x}, t_f) - 0 = \zeta^A(\mathbf{x}). \tag{3.16}$$

3.2.1 *Curvature perturbation in curvaton scenario*

Now let us apply the above to the curvaton scenario. Let us consider a stage after inflation when the inflaton has decayed to radiation, but the curvaton remains massive and behaves like dust. We label quantities associated with the radiation from the inflaton by γ and those with the curvaton by σ. Then using $P^\gamma = \rho^\gamma/3$ and $P^\sigma = 0$, Eq. (3.14) gives

$$\zeta^\gamma = \mathcal{R} + \frac{1}{4}\ln\left[\frac{\rho^\gamma(\mathbf{x},t)}{\rho^\gamma(t)}\right], \quad \zeta^\sigma = \mathcal{R} + \frac{1}{3}\ln\left[\frac{\rho^\sigma(\mathbf{x},t)}{\rho^\sigma(t)}\right], \tag{3.17}$$

where here and below we omit the fiducial position $\mathbf{x} = 0$ in the argument of fiducial quantities, e.g., $\rho(0,t) = \rho(t)$, etc. Combing these two equations, one obtains

$$\begin{aligned}
\rho^{\text{tot}}(\mathbf{x},t) &= \rho^\gamma(\mathbf{x},t) + \rho^\sigma(\mathbf{x},t) \\
&= \rho^\gamma(t)\exp[4(\zeta^\gamma - \mathcal{R})] + \rho^\sigma(t)\exp[3(\zeta^\sigma - \mathcal{R})].
\end{aligned} \tag{3.18}$$

Since what we are interested in is the curvature perturbation on the uniform total density slice in the end, let us take such a slice. Then the above becomes

$$\rho^{\text{tot}}(t) = \rho^\gamma(t) + \rho^\sigma(t)$$
$$= \rho^\gamma(t) \exp\left[4\left(\zeta^\gamma - \mathcal{R}_\rho\right)\right] + \rho^\sigma(t) \exp\left[3\left(\zeta^\sigma - \mathcal{R}_\rho\right)\right],$$
(3.19)

where $\mathcal{R}_\rho$ is the curvature perturbation on the uniform total density slices. We note that $\mathcal{R}_\rho$ is not yet conserved, though $\mathcal{R}^A_{\rho^A}\ (=\zeta^A)$ for each component is conserved. To be specific, if we linearize the equation, we find

$$\mathcal{R}_\rho(\mathbf{x},t) = \frac{4\rho^\gamma(t)\,\zeta^\gamma(\mathbf{x}) + 3\rho^\sigma(t)\,\zeta^\sigma(\mathbf{x})}{4\rho^\gamma(t) + 3\rho^\sigma(t)} + \mathcal{O}\left((\zeta^A)^2\right).$$
(3.20)

Thus it is clear that $\mathcal{R}_\rho(\mathbf{x},t)$ is not conserved.

Now we consider the decay of the curvaton. As its decay time is determined by the competition between the Hubble expansion rate and the decay rate, it is natural to consider that the decay occurs uniformly on a uniform Hubble slice in the instantaneous decay approximation. Since the uniform Hubble slicing is equivalent to the uniform total density slicing on superhorizon scales, one can then regard the decay as occurs on a uniform density slice. As soon as the decay occurs, the universe is totally dominated by radiation, and the evolution becomes adiabatic. Namely, the universe reaches the adiabatic limit instantaneously at the decay time, and the curvature perturbation $\mathcal{R}_\rho$ is conserved after that time.

For convenience, let us denote the conserved curvature perturbation on uniform density slice by ζ, that is, $\zeta(\mathbf{x}) = \mathcal{R}_\rho(\mathbf{x},t_d)$. It is determined by setting the time in Eq. (3.19) to the decay time, $t = t_d$,

$$\rho^{\text{tot}}(t_d) = \rho^\gamma(t_d) + \rho^\sigma(t_d)$$
$$= \rho^\gamma(t_d) \exp\left[4(\zeta^\gamma - \zeta)\right] + \rho^\sigma(t_d) \exp\left[3(\zeta^\sigma - \zeta)\right].$$
(3.21)

Once we know ζ^A for each component, ζ can be computed from this equation. In particular, setting $t = t_d$ in Eq. (3.20) gives the conserved curvature perturbation at linear order,

$$\zeta\left(\mathbf{x}\right)=\frac{4\rho^{\gamma}\left(t_{d}\right)\zeta^{\gamma}\left(\mathbf{x}\right)+3\rho^{\sigma}\left(t_{d}\right)\zeta^{\sigma}\left(\mathbf{x}\right)}{4\rho^{\gamma}\left(t_{d}\right)+3\rho^{\sigma}\left(t_{d}\right)}. \tag{3.22}$$

In the original curvaton scenario it is assumed that the inflaton field does not contribute to the curvature perturbation, that is, ζ^{γ} is negligible. For simplicity, we assume so here, too. Then introducing the parameter r, or equivalently $\Omega_{\sigma,d}$,

$$r\equiv\frac{3\rho^{\sigma}}{4\rho^{\gamma}+3\rho^{\sigma}}\left(t_{d}\right)=\frac{3\Omega_{\sigma,d}}{4-\Omega_{\sigma,d}}\quad\leftrightarrow\quad\Omega_{\sigma,d}=\frac{4r}{3+r}, \tag{3.23}$$

where r represents the fraction of the enthalpy density $(\rho+P)$ that the curvaton contributes at the decay time, and $\Omega_{\sigma,d}$ is that of the energy density, Eq. (3.21) becomes

$$e^{4\zeta}-\Omega_{\sigma,d}e^{3\zeta^{\sigma}}e^{\zeta}-\left(1-\Omega_{\sigma,d}\right)=0. \tag{3.24}$$

This equation can be easily solved perturbatively. Setting

$$\zeta=\sum_{n=1}^{\infty}\frac{1}{n!}\zeta_{(n)},\quad\zeta^{\sigma}=\sum_{n=1}^{\infty}\frac{1}{n!}\zeta_{(n)}^{\sigma}, \tag{3.25}$$

at linear order, we find

$$\zeta_{(1)}=\frac{3\rho^{\sigma}\left(t_{d}\right)}{4\rho^{\gamma}\left(t_{d}\right)+3\rho^{\sigma}\left(t_{d}\right)}\zeta_{(1)}^{\sigma}=r\zeta_{(1)}^{\sigma}. \tag{3.26}$$

At second order, we obtain

$$\zeta_{(2)}=r\zeta_{(2)}^{\sigma}+r\left(1-r\right)\left(3+r\right)\left(\zeta_{(1)}^{\sigma}\right)^{2}. \tag{3.27}$$

3.2.2   Spectrum and bispectrum in curvaton scenario

To compute the spectrum and bispectrum, we have to evaluate the conserved curvature perturbation on uniform curvaton density slices, ζ^{σ}. This is easily obtained by applying the δN formula to the curvaton component, which we derived in the beginning of this section,

$$\delta N^{\sigma}=\zeta^{\sigma}=\frac{1}{3}\int_{\rho^{\sigma}}^{\rho^{\sigma}+\delta\rho^{\sigma}}\frac{d\rho^{\sigma}}{\rho^{\sigma}}=\frac{1}{3}\ln\left[\frac{\rho^{\sigma}+\delta\rho^{\sigma}}{\rho^{\sigma}}\right], \tag{3.28}$$

or

$$e^{3\zeta^\sigma} = 1 + \frac{\delta\rho^\sigma}{\rho^\sigma}, \qquad (3.29)$$

where the right-hand side of the equation is to be evaluated on a flat slice at or after the onset of the oscillatory stage of the curvaton.

Then assuming that the curvaton potential can be well approximated by a quadratic potential during the oscillatory state, we have

$$\rho^\sigma = \frac{1}{2}m^2\sigma^2, \quad \delta\rho^\sigma = \frac{1}{2}m^2\left[(\sigma+\delta\sigma)^2 - \sigma^2\right]. \qquad (3.30)$$

Now we want to know the amplitude of the fluctuation $\delta\sigma$ at the onset of oscillations by relating it to the one right after the horizon crossing during inflation, say at $t = t_*$, which is Gaussian with the power per unit logarithmic interval of wavenumber $\langle\delta\sigma_*^2\rangle = H_*^2/(2\pi)^2$. If the potential is exactly quadratic, we have $\delta\sigma/\sigma = \delta\sigma_*/\sigma_*$ since the evolution equation is linear in the field value, $3H\dot\sigma = -m^2\sigma$. However, there may be some non-trivial evolution due to non-linearities.

To take this possible non-linear effect into account, it is customary to express σ as a function of σ_*,

$$\sigma = g(\sigma_*), \qquad (3.31)$$

and express $\delta\sigma$ in terms of g and its derivatives. This gives

$$\frac{\delta\rho^\sigma}{\rho^\sigma} = \frac{(\sigma+\delta\sigma)^2 - \sigma^2}{\sigma^2} = \left[1 + \frac{1}{g(\sigma_*)}\sum_{n=1}^{\infty}\frac{g^{(n)}(\sigma_*)}{n!}(\delta\sigma_*)^n\right]^2 - 1$$

$$= 2\frac{g'}{g}\delta\sigma_* + \left(\frac{g'^2}{g^2} + \frac{g''}{g}\delta\sigma_*^2\right) + \cdots. \qquad (3.32)$$

Plugging this in Eq. (3.29) and expanding the exponent, we find

$$\zeta_{(1)}^\sigma = \frac{2}{3}\frac{g'}{g}\delta\sigma_*,$$

$$\zeta_{(2)}^\sigma = \frac{2}{3}\left(-\frac{g'^2}{g^2} + \frac{g''}{g}\right)\delta\sigma_*^2 = \frac{3}{2}\left(-1 + \frac{gg''}{g'^2}\right)(\zeta_{(1)}^\sigma)^2, \qquad (3.33)$$

where $\zeta^\sigma_{(1)}$ and $\zeta^\sigma_{(2)}$ are defined by Eq. (3.25). It may be noted that $\zeta^\sigma_{(1)}$ is apparently Gaussian.

Furthermore, plugging the above expressions in Eqs. (3.26) and (3.27), we obtain ζ to second order in $\delta\sigma_*$,

$$
\begin{aligned}
\zeta &= \zeta_{(1)} + \frac{1}{2}\zeta_{(2)} \\
&= r\zeta^\sigma_{(1)} + \frac{3}{4}r\left(-1 + \frac{gg''}{g'^2}\right)(\zeta^\sigma_{(1)})^2 + \frac{1}{2}r(1-r)(3+r)(\zeta^\sigma_{(1)})^2 \quad (3.34) \\
&= r\zeta^\sigma_{(1)} + \frac{1}{2}\left[\frac{3}{2}\left(1 + \frac{gg''}{g'^2}\right) - r(2+r)\right]r(\zeta^\sigma_{(1)})^2.
\end{aligned}
$$

Identifying ζ with $\mathcal{R}_c$, denoting the Gaussian part of $\mathcal{R}_c$ by $\mathcal{R}_{c,g}$, and using the definition of f_{NL} in Eq. (2.208), we finally obtain

$$
\mathcal{R}_c = \mathcal{R}_{c,g} + \frac{3}{5}f_{NL}\left(\mathcal{R}_{c,g}\right)^2 + \cdots, \quad (3.35)
$$

where

$$
\mathcal{R}_{c,g} = r\frac{2}{3}\frac{g'}{g}\delta\sigma_*, \quad (3.36)
$$

$$
\frac{3}{5}f_{NL} = \frac{1}{2}\left[\frac{3}{2r}\left(1 + \frac{g''g}{g'^2}\right) - 2 - r\right]. \quad (3.37)
$$

Now it is straightforward to compute the spectrum and bispectrum of the curvaton model. First we consider the spectrum. From Eq. (3.36), the spectrum is given by

$$
\mathcal{P}_{\mathcal{R}_c}(k) = \frac{4\pi k^3}{(2\pi)^3}P_{\mathcal{R}_c}(k) = \left.\frac{4r^2}{9}\frac{g'^2}{g^2}\frac{H^2}{(2\pi)^2}\right|_{t_k}. \quad (3.38)
$$

If the non-linear evolution during inflation is negligible, we have

$$
\frac{g'^2}{g^2} = \frac{1}{\sigma^2(t_k)} = \exp\left[-\frac{2m^2}{3}\int_0^{N_k}\frac{dN}{H^2}\right] \approx \left(\frac{k}{k_f}\right)^{2m^2/(3H^2)}. \quad (3.39)
$$

Thus replacing ϕ with σ in the definition of η_V in Eq. (2.26), which we denote by η_σ, the spectral index in this case will be given by

$$
n_s - 1 = -2\epsilon_V + 2\eta_\sigma. \quad (3.40)
$$

The bispectrum is characterized by f_{NL}. The factor $1/r$ in front of the first term in Eq. (3.37) is simply a reflection of the fact that one needs a large intrinsic curvaton perturbation for small r to produce the observed amplitude of the curvature perturbation, and this automatically leads to a large intrinsic non-Gaussianity.

As is clear from Eq. (3.37), it will be scale independent (independent of k) if the non-linear evolution is negligible, $g'' = 0$. In this case, its amplitude is completely controlled by r. Then one immediately finds that $f_{NL} > -5/4$, and increases indefinitely as $r \to 0$. Since the latest Planck results constrain the scale-independent f_{NL} to be $f_{NL} = 2.5 \pm 5.7$ [41], one obtains the lower bound on r as

$$r \gtrsim 0.15 \,. \tag{3.41}$$

On the other hand, if $g'' \neq 0$, then f_{NL} may have a non-negligible scale dependence. In this case, the observational constraint may become very weak. Namely, a fairly large f_{NL} of $20 \sim 30$ on small k may be still allowed.

We have demonstrated how the δN formalism may be applied to the curvaton model, to compute the spectrum and bispectrum. This procedure can of course be repeated for studying higher order non-Gaussianities. In particular, it is a straightforward calculation to obtain the third order curvature perturbation. Here we simply present the expression for the third order curvature perturbation in terms of $\mathcal{R}_{c,g}$,

$$\frac{1}{3!}\zeta^{(3)} = \frac{9}{25}g_{NL}(\mathcal{R}_{c,g})^3 \,, \tag{3.42}$$

where g_{NL} is defined as the third order coefficient in ζ or $\mathcal{R}_c$ as

$$\zeta = \mathcal{R}_c = \mathcal{R}_{c,g} + \frac{3}{5}f_{NL}(\mathcal{R}_{c,g})^2 + \frac{9}{25}g_{NL}(\mathcal{R}_{c,g})^3 + \cdots, \tag{3.43}$$

and in the present case given by

$$\begin{aligned}
\frac{9}{25}g_{NL} = \frac{1}{3!}\Bigg[&\frac{9}{4r^2}\left(\frac{g^2 g'''}{g'^3} + \frac{gg''}{g'^2}\right) - \frac{9}{r}\left(1 + \frac{gg''}{g'^2}\right) \\
&+ \frac{1}{2}\left(1 - 9\frac{gg''}{g'^2}\right) + 10r + 3r^2 \Bigg].
\end{aligned} \tag{3.44}$$

The derivation of the above result is left as an exercise for the interested reader.

Application of δN formalism: Multi-brid inflation

In this chapter we study the multi-brid inflation model which is a generalization of standard hybrid inflation containing multiple inflaton fields. The motivation for studying this model is that it is simple enough such that the δN analysis can be performed analytically as we see below. On the other hand, the dynamics of the system is more non-trivial than the simple single-field models which were studied in Chapter 3 so the model can generate observable non-Gaussianity. Our presentation here follows the original papers [54] and [55].

4.1 The exact soluble class

The model we consider consists of $\mathcal{M}$ scalar fields ϕ^a, $a = 1, 2, \cdots, \mathcal{M}$, minimally coupled to gravity with the Lagrangian density

$$\mathcal{L} = \frac{M_P^2}{2} R - \frac{1}{2} g^{\mu\nu} h_{ab}(\phi) \partial_\mu \phi^a \partial_\nu \phi^b - V(\phi), \qquad (4.1)$$

where R is the Ricci scalar and h_{ab} represents the field space metric, V is the scalar fields potential and ϕ collectively represents the $\mathcal{M}$ scalar fields ϕ^a. This Lagrangian describes the dynamics of system during inflation and later on we include the so-called waterfall mechanism to terminate inflation.

The Friedmann equation and the scalar field equations respectively are given by

$$3M_P^2 H^2 = \frac{1}{2} h_{ab} \dot{\phi}^a \dot{\phi}^b + V(\phi),$$ (4.2a)

$$\ddot{\phi}^a + 3H\dot{\phi}^a + h^{ab}\partial_b V = 0.$$ (4.2b)

We are interested in the slow-roll limit in which one can neglect the kinetic energy of the scalar fields in the Friedmann equation and the second time derivative of the scalar fields can be ignored in the field equation,

$$3M_P^2 H^2 \simeq V(\phi),$$ (4.3a)

$$3H\dot{\phi}^a + h^{ab}\partial_b V \simeq 0.$$ (4.3b)

It is convenient to change the time variable from t to the number of e-folds from the end of inflation counted *backward* in time,

$$dN = -Hdt.$$ (4.4)

Note that in this convention $N > 0$ during inflation while we set $N = 0$ at the end of inflation. With this convention, the slow-roll equations of motion yield

$$\frac{d\phi^a}{dN} = \frac{h^{ab}\partial_b V}{3H^2} = M_P^2 \frac{h^{ab}\partial_b V}{V}.$$ (4.5)

Now we restrict ourselves to the models in which the slow-roll equations of motion (4.5) can be solved exactly. Specifically, they can be solved exactly if for each index a the right-hand side of Eq.(4.5) takes the form,

$$M_P^2 \frac{h^{ab}\partial_b V}{V} = \frac{f^a(\phi^a)}{F(\phi)},$$ (4.6)

in which F is an arbitrary function of the scalar fields $(\phi^1, \phi^2, ..., \phi^{\mathcal{M}})$ and f^a is a function of only ϕ^a for each a. With these assumptions, from Eqs. (4.5) and (4.6) we obtain

$$\frac{1}{f^a(\phi^a)} \frac{d\phi^a}{dN} = \frac{1}{F(\phi)}.$$ (4.7)

Introducing a new set of coordinates q^a in field space defined via

$$\ln q^a = \int \frac{d\phi^a}{f^a(\phi^a)}, \tag{4.8}$$

the equations of motion given in Eq. (4.7) become

$$\frac{d\ln q^a}{dN} = \frac{1}{F}. \tag{4.9}$$

To have a geometrical interpretation, let us introduce the radial and angular coordinates in the field space,

$$q^a = qn^a; \quad \sum_{a=1}^{\mathcal{M}} (n^a)^2 = 1. \tag{4.10}$$

Using both normalization condition for n^a and equation of motion (4.9), one can easily check that n^a is conserved,

$$\frac{dn^a}{dN} = 0, \tag{4.11}$$

while Eq.(4.9) reduces to a single equation for q in terms of N

$$\frac{d\ln q}{dN} = \frac{1}{F(q, n^a)}, \tag{4.12}$$

in which it is understood now that F is a function of q and $(n^1, n^2, \cdots, n^{\mathcal{M}})$ while each one of n^a can be determined through a constraint equation in Eq. (4.10).

We emphasize that what we have shown is not restricted to slow-roll inflation. Whenever a system approaches the attractor regime the value of the field at a given time completely determines the motion of the field at its subsequent evolution. In this attractor limit, we have $\mathcal{M}$ first-order differential equations instead of $\mathcal{M}$ second-order differential equations. In this situation, the solubility means that there are $\mathcal{M} - 1$ constants of integration. This is indeed the case we have considered above in which there are $\mathcal{M} - 1$ degrees of freedom in n^a. A schematic view of this discussion is presented in Fig. 4.1.

It is instructive to consider some specific examples. The exact solubility condition imposed in Eq. (4.6) is satisfied if one chooses a set

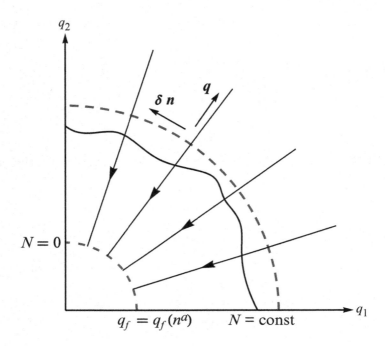

Fig 4.1 A schematic view of classical trajectories in the field space spanned by the coordinates q^a. The angular coordinates $n^a = q^a/q$ are conserved so all trajectories are radial in these coordinates. The curve indicated by $q = q_f$ denotes the surface of end of inflation which may depends on n^a. The number of e-fold N is counted backward in time from the end of inflation $q = q_f$.

of field space coordinates with the following field space metric,

$$h^{ab} = \frac{\text{diag.} \left(h^1 \left(\phi^1 \right), h^2 \left(\phi^2 \right), \cdots, h^n \left(\phi^n \right) \right)}{H \left(\phi \right)}, \tag{4.13}$$

with the potential either in the product type,

$$V = \prod_a V^a \left(\phi^a \right), \tag{4.14}$$

in which case we get

$$f^a = h^a \frac{d \ln V^a}{d\phi^a}, \quad F = H, \tag{4.15}$$

or in the form of separable type

$$V = \sum_a V^a (\phi^a), \tag{4.16}$$

in which case we have

$$f^a = h^a \frac{dV^a}{d\phi^a}, \quad F = HV. \tag{4.17}$$

Note that the separable type was first studied in details by Starobinsky [19].

Now, going back to the general case, from Eq. (4.11) we conclude that the trajectories are radial in the field space spanned by q^a and Eq. (4.12) can be solved easily for N to give

$$N(q, n^a) = \int_{q_f}^{q} F(q', n^a) \, d\ln q', \tag{4.18}$$

in which q_f is the value of q at the end of inflation. Note that in general $q_f = |\mathbf{q}_f|$ does depend on the n^a. In other words, q_f varies as the system approaches from different directions to the surface of end of inflation. Now employing the non-linear δN formula, we obtain a fully non-linear expression for the conserved comoving curvature perturbation,

$$\delta N = N(q + \delta q, n^a + \delta n^a) - N(q, n^a). \tag{4.19}$$

To develop some intuition, it is instructive to consider the linear limit of the above δN expression. Denoting its linear part by $\delta_L N$, we find

$$\delta_L N = F(q, n^a) \frac{\delta q}{q} + \int_{q_f}^{q} \frac{\partial F}{\partial n^a} \frac{dq'}{q'} \delta n^a - \frac{F}{q_f} \frac{\partial q_f}{\partial n^a} \delta n^a, \tag{4.20}$$

where (q, n^a) are calculated at the time of horizon crossing during inflation and $(\delta q, \delta n^a)$ represents the fluctuations evaluated on the flat hypersurface at that epoch. Note that δq^a are related to the fluctuations of the original field variables $\delta \phi^a$ via

$$\delta \ln q^a = \frac{\delta \phi^a}{f^a (\phi^a)} . \qquad (4.21)$$

The linear δN formula in Eq. (4.20) consists of three terms. The first term represents the contribution from the adiabatic perturbations, the second term represents the entropy perturbations during inflation [17, 56, 57], while the third term is from the entropy perturbations generated at the surface of end of inflation [58–60]. We mention that these distinctions are meaningful and useful at linear order (or perhaps perturbatively) but they are not so at fully non-linear order.

A schematic view of the above three classifications is presented in Fig. 4.2. The thick curves represent three different kinds of trajectories in the field space and the thin curves with arrows on both ends represent the field fluctuations. The dashed wavy curves represent the surface of end of inflation. After inflation ends each kind of orbit converges to a unique trajectory. The plot on the left, with the fluctuations parallel to the orbits, corresponds to the first term in Eq.(4.20) in which the fluctuations in the initial condition directly gives δN. This is the conventional adiabatic curvature perturbation in the single-field models. The plot in the middle corresponds to the second term in Eq.(4.20) in which the fluctuations are orthogonal to the orbits so they are interpreted as entropy (isocurvature) perturbations during inflation. However, by the time the two nearby orbits converge to a unique orbit, the number of e-folds depends on the route the universe has taken contributing to δN before the end of inflation. The plot on the right represents the third term in Eq. (4.20). In this case the entropy perturbations are not converted to δN until the end of inflation. However, the surface of end of inflation may not be orthogonal to the orbits giving rise to δN in the end.

4.2   Multi-brid inflation

In conventional models of inflation, curvature perturbations are generated *during* inflation. In Chapter 3 we have studied the models of single-field slow-roll inflation as main examples of this class of models.

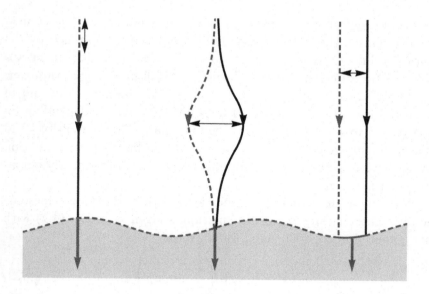

Fig 4.2 A schematic view of the three terms in Eq. (4.20). The thick curves represent three different kinds of orbits in the field space, and the thin curves with arrows on both ends represent the field fluctuations. The wavy dashed curve represents the surface of end of inflation. The left, middle and right panel, respectively, corresponds to the first term, second term, and the third term.

Here we consider a more non-trivial scenario in which curvature perturbations can be generated not only during inflation but also at the surface of end of inflation. This non-trivial example demonstrates the power of δN formalism in cosmological perturbation theory. Specializing to an extreme case, we can also consider the situation where all curvature perturbations are generated from the surface of end of inflation.

A key ingredient in this so-called "multi-brid" scenario is the role of the surface of end of inflation. The picture we have in mind is an extension of the conventional hybrid inflation [61, 62]. We assume that the heavy waterfall field χ is coupled to the inflaton fields and is locked to its local minimum during inflation. Once the inflaton field trajectory in multi-dimensional field space hits the surface of end of inflation (to be defined in the space of inflaton fields), the waterfall becomes tachyonic, terminating inflation quickly. In conventional single-field hybridinflation the point of waterfall instability is determined by the critical value of the inflaton field say $\phi = \phi_c$ and no additional curvature perturbation is generated during waterfall mechanism. However, in the multi-brid picture we are dealing with the surface of end of inflation instead of a single critical point. As a result, depending on how one approaches the surface of end of inflation in the multiple-field space, the inhomogeneities generated at the surface of end of inflation can generate additional contributions to δN and consequently curvature perturbations.

With these discussions in mind, let us study the model in some details. The Lagrangian during inflation is given as in Eq. (4.1) with the potential in the following product form

$$V(\phi) = V_0 \exp \left(\sum_a m_a \phi_a \right), \qquad (4.22)$$

in which m_a are constants with the dimension of inverse of mass. Without losing generality, we assume $m_a \phi_a > 0$ for each $a = 1, 2, ..., \mathcal{M}$. Note that the above separable potential has the advantage that the slow-roll equations can be solvedeasily enabling us to obtain an exact formula for δN. The effective mass M_a of each field ϕ_a is related to m_a via

$$M_a^2 \equiv \frac{\partial^2 V}{\partial \phi_a^2} = m_a^2 V = 3(m_a M_P)^2 H^2. \qquad (4.23)$$

Therefore, the slow-roll condition (corresponding to $M_a \ll H$) is satisfied if $m_a M_P \ll 1$.

In the slow-roll limit, the field equations are

$$\frac{d\phi_a}{dN} = \frac{1}{V} \frac{\partial V}{\partial \phi_a} = m_a. \qquad (4.24)$$

Following the geometrical logic of Eq. (4.10), we define a new set of coordinates in field space q^a given by

$$q^a = \exp \left(\frac{\phi_a}{m_a M_P^2} \right) \equiv q n^a, \qquad \sum_a (n^a)^2 = 1, \qquad (4.25)$$

which yields

$$q^2 = \sum_a \exp\left(\frac{2\phi_a}{m_a M_P^2}\right).$$ (4.26)

In terms of these new coordinates, the slow-roll equations read

$$\frac{d\ln q}{dN} = 1, \qquad \frac{dn^a}{dN} = 0.$$ (4.27)

These equations can be solved easily yielding a formula for N as a function of q: $N = \ln q + \text{const}$. Now fixing the value of N at the time of end of inflation $t = t_f$ by $N = 0$ with $q(t_f) = q_f$, we obtain

$$
\begin{aligned}
N &= \ln\left(\frac{q}{q_f}\right) \\
&= \frac{1}{2}\ln\left[\sum_a \exp\left(\frac{2\phi_a}{m_a M_P^2}\right) \Big/ \sum_a \exp\left(\frac{2\phi_{a,f}}{m_a M_P^2}\right)\right].
\end{aligned}
$$ (4.28)

The above equation enables us to find expressions for δN in terms of initial seed perturbations. However, we also need to take into account the perturbations generated from the surface of end of inflation. This means that we need to express $\phi_{a,f}$ as a function of the initial field values.

As discussed before, the picture we employ is similar to hybrid inflation in which inflation is terminated via a sharp waterfall phase transition. In this picture the waterfall field χ is very heavy during inflation so it rapidly rolls down to its local minimum $\chi = 0$. In addition, since the waterfall is very heavy, it does not contribute to large scale curvature perturbations. This claim will be discussed further in Chapter 6. Once the inflaton fields ϕ_a approach a critical surface (which we identify below), mass square of the waterfall field becomes negative signifying tachyonic instability. Afterwards, the system rolls quickly towards its global minimum and terminates inflation quickly. This can be implemented in our system if we assume

$$V_0 = \frac{1}{2}\sum_a g_a^2 \phi_a^2 \chi^2 + \frac{\lambda}{4}\left(\chi^2 - \frac{\sigma^2}{\lambda}\right)^2,$$ (4.29)

in which λ and g_a are coupling constants and σ is the global minimum of the waterfall field. The surface of end of inflation in which the waterfall instability is triggered is given by

$$\sum_a g_a^2 \phi_a^2 = \sigma^2 \,. \tag{4.30}$$

During inflation when $\sum_a g_a^2 \phi_a^2 > \sigma^2$ the waterfall is pinned in its local minimum $\chi = 0$, while after the onset of waterfall, determined in Eq. (4.30), the system rapidly goes to its global minimum $\chi = \sigma$, terminating inflation quickly. Our goal now is to determine the curvature perturbations induced from the surface of end of inflation. Note that the initial quantum fluctuations of $\delta\phi_a$ induce inhomogeneities at the surface of end of inflation Eq. (4.30) which contribute to the δN expression.

To be specific, from now on, we consider the case of two inflaton fields plus a waterfall field which terminates the inflationary period, i.e. a two-brid model. In this case, the surface of end of inflation given by Eq. (4.30) determines an ellipse in the (ϕ_1, ϕ_2) field space. It is convenient to parametrize the surface of end of inflation by the angle γ,

$$\phi_{1,f} = \frac{\sigma}{g_1} \cos\gamma, \qquad \phi_{2,f} = \frac{\sigma}{g_2} \sin\gamma \,. \tag{4.31}$$

Plugging these values of $\phi_{1,f}$ and $\phi_{2,f}$ into the expression for N given in Eq. (4.28) yields

$$N = \frac{1}{2} \ln \left\{ \left[\exp\left(\frac{2\phi_1}{m_1 M_P^2} \right) + \exp\left(\frac{2\phi_2}{m_2 M_P^2} \right) \right] \middle/ \left[\exp\left(\frac{2\sigma\cos\gamma}{g_1 m_1 M_P^2} \right) + \exp\left(\frac{2\sigma\sin\gamma}{g_2 m_2 M_P^2} \right) \right] \right\}. \tag{4.32}$$

With this prescription, the perturbations (inhomogeneities) induced at the surface of end of inflation is encoded in $\delta\gamma$. Our job is to calculate δN as a function of the initial field perturbations $\delta\phi_i$ and $\delta\gamma$. In addition, we also have to express $\delta\gamma$ as a function of initial field perturbations $\delta\phi_i$, which as a result, we obtain δN in terms of initial field fluctuations $\delta\phi_i$. Note that to calculate the power spectrum we need to find δN to linear order in field perturbations while to calculate the bispectrum we need to calculate δN to the second order. Since we are not interested in trispectrum (four-point correlation functions) and higher order correlations, in the analysis below we calculate δN to second order in field perturbations.

A very easy way to obtain γ is to solve the field equations (4.24) which yields a constraint between γ, ϕ_1 and ϕ_2. More specifically, from Eq. (4.24) we obtain

$$\phi_1 - \frac{\sigma}{g_1} \cos \gamma = m_1 M_P^2 N, \quad \phi_2 - \frac{\sigma}{g_2} \sin \gamma = m_2 M_P^2 N. \qquad (4.33)$$

Eliminating N from the above equations yields the constraint equation between γ, ϕ_1 and ϕ_2:

$$\frac{\phi_1}{m_1} - \frac{\phi_2}{m_2} = \frac{\sigma}{g_1 m_1} \cos \gamma - \frac{\sigma}{g_2 m_2} \sin \gamma. \qquad (4.34)$$

Varying the above equation to quadratic order we have

$$\frac{\delta \phi_1}{m_1} - \frac{\delta \phi_2}{m_2} = -\sigma \left(\frac{\sin \gamma}{g_1 m_1} + \frac{\cos \gamma}{g_2 m_2} \right) (\delta_1 \gamma + \delta_2 \gamma)$$
$$- \frac{\sigma}{2} \left(\frac{\cos \gamma}{g_1 m_1} - \frac{\sin \gamma}{g_2 m_2} \right) (\delta_1 \gamma)^2, \qquad (4.35)$$

in which $\delta_1 \gamma$ and $\delta_2 \gamma$ represent respectively the linear and the quadratic perturbations in γ. It is important to note that γ is a non-linear function of the fields so we have to calculate the intrinsic second order perturbations $\delta_2 \gamma$. Solving the above equations order by order yields the relations between $\delta \gamma$ and the field perturbations as follows:

$$\delta_1 \gamma = -\frac{g_1 g_2}{\sigma} \frac{m_2 \delta \phi_1 - m_1 \delta \phi_2}{g_1 m_1 \cos \gamma + g_2 m_2 \sin \gamma}, \qquad (4.36)$$

and

$$\delta_2 \gamma = \frac{1}{2} \frac{g_1 m_1 \sin \gamma - g_2 m_2 \cos \gamma}{g_1 m_1 \cos \gamma + g_2 m_2 \sin \gamma} (\delta_1 \gamma)^2$$
$$= \frac{g_1^2 g_2^2}{2\sigma^2} \frac{g_1 m_1 \sin \gamma - g_2 m_2 \cos \gamma}{(g_1 m_1 \cos \gamma + g_2 m_2 \sin \gamma)^3} (m_2 \delta \phi_1 - m_1 \delta \phi_2)^2. \qquad (4.37)$$

Having calculated $\delta\gamma$, it is straightforward to calculate δN by perturbing Eq. (4.32) to second order. However, an easier way is to vary either equations in Eq. (4.33) which yields

$$
\delta N = \frac{g_1 \cos\gamma\,\delta\phi_1 + g_2 \sin\gamma\,\delta\phi_2}{M_P^2\left(g_1 m_1 \cos\gamma + g_2 m_2 \sin\gamma\right)}
$$
$$
+ \frac{g_1^2 g_2^2}{2 M_P^2 \sigma}\frac{\left(m_2 \delta\phi_1 - m_1 \delta\phi_2\right)^2}{\left(m_1 g_1 \cos\gamma + m_2 g_2 \sin\gamma\right)^3}.
$$

(4.38)

This completes our job to find δN to second order in terms of the initial field fluctuations.

Before going to the calculation of the power spectrum, let us make a comment about the surface of the end of inflation. To be rigorous, there is a correction to be added to the above formula. It comes from the fact that the surface where the inflation ends in the field space, Eq. (4.30), is not a surface of constant energy density. However, one can show that for small m_1 and m_2 irrespective of the values of σ, g_1 and g_2 this correction can be ignored [54].

4.3 The power spectrum and the bispectrum in multi-brid scenario

Having obtained δN to the second order in field perturbations, we are ready to calculate the power spectrum and the bispectrum.

To start with, we need the statistics of the initial field fluctuations $\delta\phi_a$. We know they originate from quantum fluctuations deep inside the horizon, i. e. when $k \gg aH$ for the momentum mode k. The initial information about the statistics of $\delta\phi_a$ are encoded once they leave the horizon when $k = aH$ after which the modes become classical. We impose the simple Bunch–Davies (Minkowski) initial conditions in which the field fluctuations are random Gaussian fields obeying the following simple correlations,

$$
\frac{k^3}{2\pi^2}\langle \delta\phi_a \delta\phi_b \rangle = \left(\frac{H}{2\pi}\right)^2 \Bigg|_{t_k} \delta_{ab},
$$

(4.39)

in which t_k indicates that the expectation is calculated at the time of horizon crossing.

With this prescription for the expectation value of field products at the time of horizon crossing, the power spectrum of curvature perturbation can be found as

$$\mathcal{P}_{\mathcal{R}_c} = \frac{k^3}{2\pi^2} P_{\mathcal{R}_c}(k) = \frac{g_1^2 \cos^2 \gamma + g_2^2 \sin^2 \gamma}{M_P^2 (m_1 g_1 \cos \gamma + m_2 g_2 \sin \gamma)^2} \left(\frac{H}{2\pi M_P} \right)^2.$$
(4.40)

Correspondingly, the spectral index is obtained to be

$$n_s = 1 - (m_1^2 + m_2^2).$$
(4.41)

In addition, the ratio of the tensor power spectrum to scalar power spectrum r is given by

$$r = 8 \frac{(m_1 g_1 \cos \gamma + m_2 g_2 \sin \gamma)^2}{g_1^2 \cos^2 \gamma + g_2^2 \sin^2 \gamma}.$$
(4.42)

Straightforward calculations of the non-Gaussianities using δN formalism show the advantage of this powerful method. δN analysis significantly reduces the amount of calculations required for finding non-Gaussianity parameter f_{NL}. With two scalar field fluctuations $\delta \phi_1$ and $\delta \phi_2$ it is convenient to use the notion of the adiabatic and the entropic perturbations, in which the adiabatic perturbation is simply the curvature perturbations $\mathcal{R}_c$ while the entropy perturbation S is defined such that $\langle RS \rangle = 0$. With some efforts, one can check that Eq. (4.38) can be written as

$$\delta N = \mathcal{R}_L + \frac{3}{5} f_{NL} (\mathcal{R}_L + S)^2,$$
(4.43)

in which the linear part of the curvature perturbation is

$$\mathcal{R}_L = \frac{g_1 \cos \gamma \delta \phi_1 + g_2 \sin \gamma \delta \phi_2}{M_P^2 (g_1 m_1 \cos \gamma + g_2 m_2 \sin \gamma)},$$
(4.44)

while the entropy perturbation is

$$S = \frac{g_2 \sin \gamma \delta \phi_1 - g_1 \cos \gamma \delta \phi_2}{M_P^2 (g_1 m_2 \cos \gamma - g_2 m_1 \sin \gamma)}.$$
(4.45)

Interestingly, Eq. (4.43) has the form of local-type non-Gaussianity with the non-Gaussianity parameter f_{NL}^{local} given by

$$f_{NL}^{\text{local}} = \frac{5g_1^2 g_2^2}{6(g_1^2 \cos^2 \gamma + g_2^2 \sin^2 \gamma)^2} \frac{M_P^2 (g_1 m_2 \cos \gamma - g_2 m_1 \sin \gamma)^2}{\sigma (g_1 m_1 \cos \gamma + g_2 m_2 \sin \gamma)}. \tag{4.46}$$

In order for the model to be consistent with the cosmological observations, such as the Planck observations, we require $n_s \simeq 0.96, r < 0.1$ and $|f_{NL}^{\text{local}}| \lesssim 5$. As can be seen from the above expressions for n_s, r and f_{NL}, the multi-brid model has enough free parameters such as m_1, m_2, g_1 and g_2 that one can easily satisfy the above observational constraints.

Having presented the δN analysis for the curvature perturbations and non-Gaussianity, one may ask what is the physical reason for generating non-trivial curvature perturbations and non-Gaussianity? Intuitively speaking, δN has contributions not only during inflation but also contributions from the inhomogeneities generated at the surface of end of inflation. We have a two-field system with a given surface of end of inflation, i.e. Eq. (4.30). The initial quantum fluctuations $\delta \phi_1$ and $\delta \phi_2$ generate inhomogeneities at the surface of end of inflation $\delta \phi_{1,f}$ and $\delta \phi_{2,f}$. Depending on how one approaches the surface of end of inflation, as determined by the angle γ, non-trivial curvature perturbations and non-Gaussianity can be generated from the inhomogeneities generated at the surface of end of inflation as calculated in details above.

As an interesting extreme example, consider the case in which only one field is responsible for inflation, say ϕ_1, while both fields determine the surface of end of inflation. In this view, ϕ_2 is an isocurvature with $m_2 \simeq 0$ which can induce inhomogeneities only by modulating the surface of end of inflation [58–60]. In this limit, from Eq. (4.40) we obtain

$$\mathcal{P}_{\mathcal{R}_c} = \left(1 + \frac{g_2^2}{g_1^2} \tan^2 \gamma\right) \left(\frac{H}{2\pi m_1 M_P^2}\right)^2 \qquad (m_2 = 0). \tag{4.47}$$

Note that the modulation effect induced by the isocurvature field ϕ_2 is captured by the $\tan \gamma$ term above. In the absence of the isocurvature field we recover the usual formula $\mathcal{P}_{\mathcal{R}_c} = \frac{H^2}{8\pi^2 \epsilon M_P^2}$ by noting that the slow-roll parameter ϵ is given by $\epsilon = m_1^2 M_P^2/2$.

In the extreme case where all perturbations are generated from the inhomogeneities induced at the surface of end of inflation, as envisaged by Lyth [60], we have

$$\mathcal{P}_{\mathcal{R}_c} = \frac{g_2^4 \phi_2^2}{g_1^4 \phi_{1,f}^2} \left(\frac{H}{2\pi m_1 M_P^2} \right)^2 \qquad \left(m_2 = 0, \quad \frac{g_2^4 \phi_2^2}{g_1^4 \phi_{1,f}^2} \gg 1 \right).$$

(4.48)

Note that since ϕ_2 is an isocurvature field, it does not roll during inflation so there is no difference between $\phi_{2,f}$ and the value of ϕ_2 during inflation.

Application of δN formalism: Non-attractor inflation

5.1 Motivation for non-attractor inflation

In this chapter we present the application of δN formalism in non-attractor inflationary backgrounds which is particularly interesting since it could shed lights on the strength of δN formalism. Before presenting our setup let us provide the motivation for studying the non-attractor scenarios.

One great advantage of many models of inflation is that inflation is insensitive to initial conditions. For example, consider the simple chaotic inflationary models with the potential $V = m^2\phi^2/2$. In the two-dimensional phase space of $(\phi, \dot\phi)$, if one starts with sufficiently large initial value of ϕ, say $\phi \gtrsim 10M_P$, then the inflation is an attractor solution. In other words, independent of the initial velocityof the field $\dot\phi$, the system rapidly approaches the slow-roll inflation regime. The friction term from the background expansion, $3H\dot\phi$, slows down the system and it shortly approaches the slow-roll regime. Whether or not one gets enough number of e-folds to solve the flatness and the horizon problem depends on if one starts with large enough initial value of the field in the phase space.

The common picture for simple models of inflation is that the inflaton field rolls down from the top of the potential and quickly enters the attractor phase as described above. However, in the generic field space one can consider the situation where the inflaton field moves towards the top of the potential. Intuitively speaking this is the situation discussed in this chapter. One can show that when the inflaton field moves up the potential, $\dot\phi$ decreases exponentially. As a result, one cannot neglect $\ddot\phi$ in this period. It will be illustrative to translate this situation in terms of the slow-roll parameters $\epsilon = -\dot{H}/H^2 = \dot\phi^2/2M_P^2$ and $\eta = \dot\epsilon/\epsilon H$. When $\dot\phi$ is exponentially decaying, ϵ decays exponentially too and as a result one can no longer assume that $\eta \ll 1$. This is unlike the common case, in which one assumes both ϵ and η are small in order to obtain a scale-invariant power spectrum with $n_s \simeq 1$. As we will see, one can consider the limit of non-attractor solution in which a scale-invariant power spectrum can be obtained when $\eta \simeq -6$.

One interesting feature of non-attractor models is that the curvature perturbation on superhorizon scales is not conserved [63–66]. This is in contrast to the usual conclusion that for single-field models $\mathcal{R}_c$ is constant on superhorizon scales. This will have non-trivial implications for cosmological observations. One implication which we will study in details in this chapter is Maldacena's consistency condition for the local-type non-Gaussianity in the squeezed limit for single-field models with vacuum initial condition [44,67]. Observationally, the single-field consistency condition implies that all single-field models of inflation are ruled out if $f_{NL}^{\rm local} \approx 1$ or larger are detected. As we shall see, the Maldacena's consistency condition assumes that $\mathcal{R}_c$ is constant on superhorizon scales. However, in non-attractor models in which $\mathcal{R}_c$ evolves on superhorizon scales, the consistency condition can be violated. As a result, a detection of $f_{NL}^{\rm local}$ in the local-shape does not necessarily ruled out *all* single-field models.

With these general discussions in mind, in Sec. 5.2 we present the non-attractor models in the general single-field models and then calculate δN to second orders in field perturbations. A crucial aspect of δN formalism appearing in these analyses is that N should be considered as a function of both ϕ and $\dot\phi$ in the two-dimensional $(\phi, \dot\phi)$ phase space. This is in contrast to the usual solution in which the system has reached the attractor phase and one needs to consider N just as a function of ϕ itself on the initial flat hypersurface.

5.2   Non-attractor background

In this section we present our non-attractor model following closely Ref. [65] (see also Refs. [63,64,68]). While we refer to this setup as non-attractor model, but alternatively this setup is also referred to as "ultra slow-roll" in literature [69,70].

We mention that the model requires several fine tunings both on initial conditions and on model parameters for the purpose of the analysis below. The fact that the model suffers from the fine tunings is not an issue for us in our discussions below. The reason is that our goal is to present a counter example for the single-field model in which the constancy of $\mathcal{R}_c$ on superhorizon scales is violated. As a result, the Maldacena's consistency condition is violated. This is important since the Maldacena's consistency condition is usually considered as a fundamental theorem, valid for "all" single-field models with the *vacuum* initial conditions.

The model we consider is a scalar field with a non-standard kinetic energy such as in models of k-inflation [71,72] coupled minimally to gravity,

$$ S = \int dt \, d^3x \sqrt{-g} \left[\frac{M_P^2}{2} R + P(X, \phi) \right], \tag{5.1} $$

in which $X \equiv -\frac{1}{2} \partial_\mu \phi \partial^\mu \phi$, and

$$ P(X, \phi) = X + \frac{X^\alpha}{M^{4\alpha-4}} - V(\phi), \qquad V(\phi) = V_0 + v \left(\frac{\phi}{M_P} \right)^\beta. \tag{5.2} $$

Here M, α, v, V_0, and β are free constant parameters. We take M, α, V_0, and β to be positive while we require $v < 0$ for non-attractor solution as we shall see below.

Before we proceed to our non-attractor analysis, we should mention that the models based on $P(X, \phi)$ theory are vastly studied in the literature. In particular, it is known that these models can generate large equilateral-type non-Gaussianity if the sound speed of the perturbations, c_s, is small, such as in model of DBI inflation [48,73], for a review see Ref. [43].

We consider the initial conditions for $\dot{\phi}$ such that the inflation has two stages. During the first stage the inflaton field climbs up the potential and the term linear in X in $P(X,\phi)$ is sub-dominant during this phase. This corresponds to our non-attractor phase as described above. Towards the end of the non-attractor phase when the kinetic energy has decayed sufficiently, the term linear in X dominates and the second phase of inflation begins. Furthermore, we assume that the second phase is a usual slow-roll inflation and inflation terminates once the slow-roll conditions are violated. As we shall see the crucial point is that the curvature perturbation is not conserved on superhorizon scales during the early non-attractor phase while it is conserved on superhorizon scales for the modes which leave the horizon during the attractor phase.

The background field equations of motion are

$$3M_{\rm P}^2 H^2 = 2XP_{,X} - P, \qquad (5.3a)$$

$$M_{\rm P}^2 \dot{H} = -XP_{,X}, \qquad (5.3b)$$

and

$$\left(P_{,X} + 2XP_{,XX}\right)\ddot{\phi} + 3H\dot{\phi}P_{,X} + 2XP_{,X\phi} - P_{,\phi} = 0, \qquad (5.3c)$$

in which the dot denotes the derivative with respect to the cosmic time, t, and we use the convention that $P_{,X} \equiv \partial P/\partial X$ and so on. As in slow-roll models of inflation we assume that the constant term in the potential dominates in the total energy density and

$$3M_{\rm P}^2 H^2 \simeq V_0. \qquad (5.4)$$

The sound speed, c_s, and slow-roll parameters, ϵ and η, are given by

$$c_s^2 \equiv \frac{P_{,X}}{P_{,X} + 2XP_{,XX}}, \qquad (5.5a)$$

$$\epsilon \equiv -\frac{\dot{H}}{H^2} = \frac{XP_{,X}}{M_{\rm P}^2 H^2}, \qquad (5.5b)$$

$$\eta \equiv \frac{\dot{\epsilon}}{H\epsilon} = \frac{\ddot{\phi}}{H\dot{\phi}}\left(1 + \frac{1}{c_s^2}\right) + \frac{\dot{\phi}P_{,X\phi}}{HP_{,X}} + 2\epsilon. \qquad (5.5c)$$

First we consider the non-attractor phase in which the term linear in X in $P(X,\phi)$ can be neglected. In order to be able to perform the analytic calculations, we assume that c_s and η are nearly constant.

The latter condition implies that $\epsilon \propto a^\eta$. We will check below that our Lagrangian can satisfy these conditions. With these assumptions, the sound speed during the non-attractor phase is given by

$$c_s^2 \simeq \frac{1}{2\alpha - 1}. \tag{5.6}$$

Since $P_{,X\phi} = 0$ in our model, the Klein–Gordon equation, Eq. (5.3c), can be rewritten as

$$\frac{P_{,X}}{c_s^2}\dot{X} + 6HXP_{,X} - P_{,\phi}\dot{\phi} = 0. \tag{5.7}$$

It is not easy to find the analytical solutions of the above equation. Instead, we consider the following ansatz

$$\phi(t) = \text{const} \times a^\kappa, \tag{5.8}$$

where κ is a constant which will be determined from the consistency of the solutions. From this ansatz, and noting that H is nearly constant, we obtain

$$\dot{\phi} \simeq H\kappa\phi, \qquad \ddot{\phi} \simeq H^2\kappa^2\phi, \qquad X \simeq \frac{1}{2}H^2\kappa^2\phi^2. \tag{5.9}$$

Plugging the above relations into Eq. (5.7), one obtains the following equation:

$$2\alpha M^4 \left(\frac{V_0\kappa^2}{6M^4}\right)^\alpha \left(\frac{\phi}{M_p}\right)^{2\alpha}\left(1 + \frac{3c_s^2}{\kappa}\right) + v\beta c_s^2 \left(\frac{\phi}{M_p}\right)^\beta \simeq 0. \tag{5.10}$$

From the above, we obtain two equations, namely one equation for the matching of the powers of ϕ and the other equation for the cancellation of constant pre-factors. These two consistency equations yield

$$\beta = 2\alpha = \frac{1}{c_s^2} + 1, \tag{5.11}$$

$$v = -\frac{M^4}{c_s^2}\left(\frac{V_0\kappa^2}{6M^4}\right)^\alpha \left(1 + \frac{3c_s^2}{\kappa}\right). \tag{5.12}$$

This indicates that $v < 0$. In addition, we also have

$$\epsilon = \frac{XP_{,X}}{M_P^2 H^2} \propto a^{2\alpha\kappa}. \tag{5.13}$$

Recalling that $\epsilon \propto a^{\eta}$ we obtain

$$\kappa \simeq \frac{\eta}{2\alpha}. \tag{5.14}$$

This also indicates that η is nearly constant. In addition, note that in order for the field to climb up the potential we need $\dot{\phi}$ to have a sign opposite of ϕ so $\kappa < 0$. This is consistent with the expectation that ϵ decays exponentially during the non-attractor phase, as can be seen from Eq. (5.13).

We have five free parameters in our model, as given in $P(X, \phi)$ in Eq. (5.2). In addition, κ is another parameter appearing in the solution. Among all, two of them are determined by requiring that the ansatz (5.8) to be a consistent solution and two other parameters are fixed for a given value of η and c_s. At the end, we are left with two undetermined parameters. As we shall see later, requiring a scale-invariant power spectrum enforces $\eta \simeq -6$ while c_s is fixed by the level of non-Gaussianity predicted by the model.

Using Eqs. (5.9) and (5.11), one can show that

$$\frac{X^{\alpha}/M^{4\alpha-4}}{v(\phi/M_P)^{\beta}} \simeq -c_s^2 \left(1 + \frac{3c_s^2}{\kappa}\right)^{-1}. \tag{5.15}$$

Therefore, for $c_s < 1$, the kinetic term is always sub-dominant in comparison to the potential term.

In the above calculations, we have assumed that the term linear in X is sub-dominant, i.e., $X \ll X^{\alpha}/M^{4\alpha-4}$, or equivalently $(X/M^4)^{\alpha-1} \gg 1$. For $\alpha \gg 1$ corresponding to $c_s \ll 1$, this implies that $X/M^4 > 1$. Using the ansatz (5.8), this condition translates into

$$\sqrt{\frac{V_0}{6M_P^2}} \frac{|\kappa|\phi}{M^2} > 1. \tag{5.16}$$

This condition breaks down at $\phi = \phi_*$ or $t = t_*$ defined by

$$\frac{\phi_*}{M_P} \simeq \sqrt{\frac{6}{V_0}} \frac{M^2}{|\kappa|}. \tag{5.17}$$

After ϕ_*, the system enters the slow-roll inflation phase for a relatively wide range of initial conditions. If this does not happen, we loose our analytic control on the solution and the curvature perturbation may not be conserved in the second phase. Therefore, in our analysis below, we assume that the second inflationary phase does occur and is in the slow-roll regime.

In our picture of the model presented above, the field climbs up the potential during the first stage of inflation. This is why we have a non-attractor background initially. Note that the ansatz given in Eq. (5.8) and the fine-tuning associated with the parameters given in Eqs. (5.11), (5.12) and (5.14) are *not* the necessary conditions for obtaining the general non-attractor solution. We require these specific values of the parameters in order to be able to perform the analytic calculations for the specific solution given in Eq. (5.8). We mention that one can find a non-attractor phase for half of the ranges of possible initial conditions.

Depending on the initial conditions, the system admits three different solutions: the undershoot, the critical or the overshoot. In the undershoot solution, the inflaton field climbs up the potential, stops somewhere before reaching the origin (the top of the potential), turns around and rolls down on the same side of the potential. In this case ϕ always has a unique sign while $\dot{\phi}$ changes the sign, see Fig. 5.1. In the overshoot solution, the inflaton field climbs up the potential with a large enough initial velocity, so that it goes over the top of the potential, and rolls down from the other side of the potential. In this case the sign of ϕ changes while $\dot{\phi}$ always has a unique sign, see Fig. 5.2. The critical solution is the limit when the initial conditions are such that it takes infinite amount of time for the inflaton field to reach the top of the potential. Of course in this limit, inflation never end so one has to employ additional mechanism to terminate the non-attractor phase of inflation, see Ref. [63] for an example.

The three different solutions of the inflaton-field evolution in the phase space are shown in Fig. 5.3 The critical solution (the black solid curve in Fig. 5.3) separates the overshoot and undershoot solutions. The early-time behavior of this curve for large $|\phi|$ and $|\dot{\phi}|$, in which the term X^α dominates, is asymptotically the same as the ansatz we obtained above. However, the linear term in X dominates the $P(X)$ near the origin and one has to solve the equation of motion for a

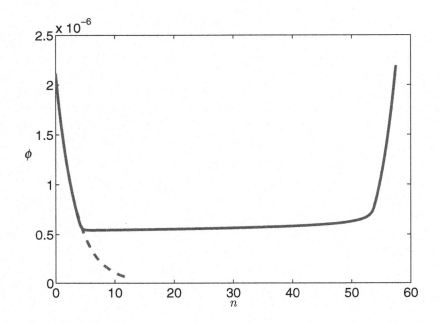

Fig 5.1 Here the evolution of $\phi(n)$ in the undershoot situation as a function of the number of e-folds, n, counted from the beginning of inflation is presented. The inflaton field climbs up the potential, stops somewhere before reaching the top of the potential $\phi = 0$, turns around and goes back to $+\infty$. The dashed red curve represents the analytic ansatz for the non-attractor phase, while the solid blue curve is the full numerical solution. The transition to the second stage slow-roll inflation phase is sharp and an extended slow-roll phase follows afterward. The parameters are $V_0 \simeq 6.25 \times 10^{-4} M_P^4$, $M = 5 \times 10^{-5} M_P$, $a = 10$, and $\eta = -6$. The figure is borrowed from Ref. [65].

canonically normalized field given by

$$\ddot{\phi} + 3H\dot{\phi} + \frac{v\beta}{M_P}\left(\frac{\phi}{M_P}\right)^{\beta-1} = 0. \qquad (5.18)$$

For $\beta \gg 1$, the last term proportional to $(\phi/M_P)^{\beta-1}$ is small compared to the first two terms near the origin in which $\phi/M_P \simeq 0$. Thus, the slow-roll condition no longer holds, and we have $\ddot{\phi} + 3H\dot{\phi} \simeq 0$. This

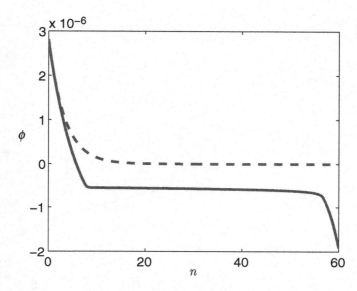

Fig 5.2 This plot is the same as Fig. 5.1, but for the overshoot situation. The inflaton field climbs up the potential, goes over the top of the potential, and rolls down from the other side of the potential. The figure is borrowed from Ref. [65].

corresponds to the scenario of non-attractor inflation with a constant potential studied in Ref. [63]. As a result $d\phi/dn \simeq e^{-3n} \propto a^{-3}$ in which n is the number of e-folds counted from the beginning of inflation. As a result $\epsilon \propto e^{-6n} \propto a^{-6}$ and $\eta \simeq -6$. This asymptotic solution is in agreement with the numerical one for the region in phase space where $\phi/M_{\rm P} \simeq 0$, see the purple dashed-line line in the middle of Fig. 5.3.

5.3   Power spectrum for non-attractor background

In this section we calculate the power spectrum of curvature perturbations generated during the non-attractor phase and study the condition for obtaining a scale-invariant power spectrum. In addition, these analyses are also needed in order to calculate the seed initial quantum fluctuations at the time of horizon crossing for the subsequent δN analysis.

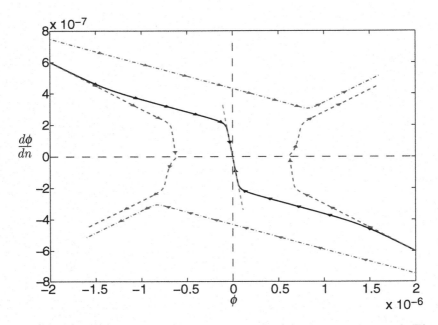

Fig 5.3 Here the phase-space diagram of the model is presented. The black solid curve separates two different trajectories: the undershoot (dashed blue curves) and overshoot (dot-dashed red curves) trajectories. The purple dashed line near the origin of the plot indicates the asymptotic solution, $\phi \propto a^{-3}$. In the undershoot case, the inflaton field climbs up the potential, stops somewhere before reaching the top of the potential, and returns back on the same side of the potential. In the overshoot case, the field climbs up the potential, crosses the top of the potential, and rolls down from the other side of the potential. The symmetry in the plot represents the fact that our Lagrangian given in Eq. (5.2) is symmetric under the transformation $\phi \to -\phi$ and $\dot{\phi} \to -\dot{\phi}$. The figure is borrowed from Ref. [65].

In order to obtain the seed quantum fluctuations, one has to find the second order action for the metric and scalar field fluctuations. This is a standard procedure for the simple canonically normalized scalar field theories. For the second order action in our model, with the non-canonical kinetic energy given by the action (5.1), see Refs.

[48,72,74]. The quadratic action for the curvature perturbation on comoving surface $\mathcal{R}_c$ is

$$S = \frac{1}{2} \int d^3x d\tau \, z^2 \left[\mathcal{R}_c'^2 - c_s^2 (\nabla \mathcal{R}_c)^2 \right], \tag{5.19}$$

in which the prime denotes the derivative with respect to the conformal time τ and

$$z^2 \equiv \frac{2\epsilon a^2}{c_s^2} M_P^2. \tag{5.20}$$

The equation of motion obtained from the above action is

$$\left(\mathcal{R}_{c_k}' \epsilon a^2 \right)' + c_s^2 k^2 \epsilon a^2 \mathcal{R}_{c_k} = 0. \tag{5.21}$$

Below we solve the above equation for all Fourier modes. However, in order to get some insights for the unusual behavior of the superhorizon modes in non-attractor models, let us consider the superhorizon limit of the above equation corresponding to $c_s k \to 0$. In this limit, the above equation can be solved easily yielding

$$\mathcal{R}_c = C_1 + C_2 \int \frac{d\tau}{\epsilon a^2}, \tag{5.22}$$

in which C_1 and C_2 are two constants of integration. In conventional slow-roll models in which ϵ is nearly constant, the term containing C_2 in the above equation decays exponentially during inflation and one recovers the celebrated result that on superhorizon scales $\mathcal{R}_c$ is constant. However, in our non-attractor phase when ϵ decays exponentially, one may not neglect the term containing C_2 in the above equation. Indeed, with $\epsilon \propto a^{-6}$ and $\eta \simeq -6$ (as we shall see below) we see that the term containing C_2 grows rapidly and will dominate over the constant term C_1. This is the essence of the non-attractor phase in which the curvature perturbation is not frozen on superhorizon scales.

Now we solve Eq. (5.21) for the general case recalling that $\epsilon \propto a^\eta$. The general solution is given in terms of Hankel functions $H_\nu^{(1)}(x)$ and $H_\nu^{(2)}(x)$ corresponding respectively to the positive and negative frequency modes. Imposing the Bunch–Davies or Minkowski initial state deep inside the horizon, the $H_\nu^{(1)}(x)$ solution survives and we have

$$\mathcal{R}_{c_k} = C x^\nu H_\nu^{(1)}(x) , \qquad (5.23)$$

where we have defined

$$x \equiv -c_s k\tau, \qquad \nu \equiv \frac{3+\eta}{2}, \qquad (5.24)$$

and C is a constant given by

$$|C|^2 \equiv \frac{\pi c_s}{8 k \epsilon_i a_i^2 M_P^2} x_i^{1-2\nu} . \qquad (5.25)$$

The subscripts i denote the corresponding values at the start of inflation (or at the start of non-attractor phase).

As we shall see below, in order to get a nearly scale-invariant power spectrum we require $\eta \simeq -6$. As a result, $\nu \simeq -3/2$. This is in contrast to usual the slow-roll case in which $\nu \simeq 3/2$. Now, using the small argument limit of the Hankel function for $\nu \simeq -3/2$, we see that $\mathcal{R}_{c_k} \propto (-k\tau)^{-3} \propto a^3$. As a result, the curvature perturbation grows exponentially during the non-attractor phase. In order to prevent the system to become unstable, the non-attractor phase is to be terminated at $\phi = \phi_*$. In the overshoot and undershoot solutions this happens dynamically in our model. However, for the critical case in which it takes infinite time for the inflaton field to reach the top of the potential, the non-attractor phase never ends and one has to terminate it by hand as in Ref. [63]. Note that after the non-attractor phase and during the slow-roll phase $\mathcal{R}_c$ becomes a constant. Therefore, one has to follow the evolution of curvature perturbation only till the end of non-attractor phase at $t = t_*$ when $\phi = \phi_*$ given in Eq. (5.17).

The power spectrum of curvature perturbations at the end of the non-attractor phase is given by the usual formula $\mathcal{P}_\mathcal{R} = \frac{k^3}{2\pi^2} |\mathcal{R}_{c_k}|^2$. Using Eq. (5.25) one can write the power spectrum in terms of the parameters at the end of the non-attractor phase as

$$\mathcal{P}_\mathcal{R} \simeq \frac{\Gamma(|\nu|)^2}{\pi^3 2^{2\nu+4}} \left(\frac{H_*}{M_P}\right)^2 \frac{1}{c_s \epsilon_*} \left(\frac{c_s k}{H_* a_*}\right)^{3+2\nu} . \qquad (5.26)$$

We have assumed that $\nu = (3 + \eta)/2 < 0$ so one can expand the Hankel function for the small argument limit, $x \ll 1$.

We emphasis again that during the non-attractor phase the fast decay of ϵ causes the curvature perturbation to grow very rapidly on superhorizon scales. This growth continues until the end of the non-attractor phase at $t = t_*$ or $\phi = \phi_*$ given in Eq. (5.17). The curvature perturbation is conserved during the subsequent slow-roll phase. Therefore, $\mathcal{P}_{\mathcal{R}_c}$ calculated at $t = t_*$ is the observable power spectrum for themodes which the horizon during the non-attractor phase.

With the power spectrum given in Eq. (5.26), the spectral index is given by

$$n_s - 1 \simeq 3 + 2\nu = 6 + \eta. \tag{5.27}$$

Therefore, $\eta = -6$ is required in order to obtain scale-invariant power spectrum. A slightly red-tilted power spectrum, $n_s = 0.96$, as preferred by the cosmological observations, can be easily obtained by choosing $\eta = -6.04$.

5.4   δN formalism in non-attractor backgrounds

Having presented our setup and solving the background dynamics, we are ready to use δN formalism for the non-attractor backgrounds. However, extra care must be taken when dealing with δN formalism in non-attractor backgrounds. In the usual case once the system enters the slow-roll solutions, ϕ plays the role of aclock and from the first order equation $3H\dot{\phi} + V_\phi \simeq 0$ $\dot{\phi}$ is uniquely determined as a function of ϕ. As a result, N is just a function of ϕ, so when we calculate δN we need to consider only the perturbations of the scalar-field trajectories with respect to the field's value ϕ_i at the initial hypersurface. However, in the non-attractor case we also need the information of $\dot{\phi}$ to determine the background trajectory. In other words, we have to solve N in the two-dimensional phase space of $(\phi, \dot{\phi})$. As a result, the number of e-folds N counted backward from the epoch when $\phi = \phi_*$ to an earlier epoch is a function of both ϕ and $\dot{\phi}$, $N = N(\phi, \dot{\phi}; \phi_*)$.

During the subsequent attractor phase, we assume as usual that the evolution of the universe is unique irrespective of the value of its velocity $\dot{\phi}_*$, so at and after $t = t_*$, the scalar field plays the role of a clock. This is a necessary condition for the applicability of the usual δN formalism, since only in this case δN is equal to the final value of the comoving curvature perturbation $\mathcal{R}_c$ which is conserved at $t \geq t_*$.

With this description in mind, we apply the δN formalism to our model. Our program is as follows. In order to find the background scalar-field trajectories $N = N(\phi, \dot{\phi})$, we solve the equation of motion of the scalar field perturbatively by expanding it around the particular trajectory given by $\phi \propto e^{\kappa H t}$. We then use these background solutions for the field trajectories to compute the perturbations of the number of e-folds with respect to the initial field values of ϕ and $\dot{\phi}$.

It turns out that the analysis for the case $c_s \neq 1$ is somewhat non-trivial. In order to get insights as how δN works in this setup before getting involved in the technicalities associated with the case $c_s \neq 1$, first we consider thesimple case $c_s = 1$ so $P(X, \phi)$ is a linear function of X as in standard canonically normalized models of inflation. We relegate the analysis for the case of arbitrary c_s into Appendix A.

5.4.1   *The case with $c_s = 1$*

To familiarize ourselves with the δN calculation in the non-attractor background, let us first work out the simplest case with the canonical kinetic term, $c_s = 1$ and $V = V_0$, so $P = X - V_0$.

During the non-attractor phase the background Klein–Gordon equation is given by

$$\ddot{\phi} + 3H\dot{\phi} = 0. \tag{5.28}$$

This equation can be solved easily yielding

$$\phi = \lambda + \mu e^{-3Ht}, \tag{5.29}$$

where λ and μ are constants of integration and we have assumed $\dot{\phi} > 0$ without loss of generality. From this solution one easily obtains $\epsilon \propto \dot{\phi}^2 \propto a^{-6}$ and $\eta = -6$. The fact that ϵ decays exponentially indicates that the kinetic energy is decaying rapidly and the system approaches a pure dS background dominated by the constant potential term V_0. As a result inflation does not terminate in this setup. In order to terminate inflation, one can couple ϕ to a heavy waterfall field, so once

ϕ reaches ϕ_* a waterfall mechanism turns on and the non-attractor phase quickly ends and we enter the second slow-roll phase. Note that the waterfall is very heavy during inflation so it does not affect the curvature perturbations on superhorizon scales. Alternatively, one can follow the method employed in Ref. [63] and glue the constant potential term to a varying potential $V = V_>(\phi)$ for $\phi > \phi_*$ such that $V_>(\phi)$ can support a period of slow-roll inflation.

The number of e-folds counted backward from the time $t = t_*$ is

$$N = \int_t^{t_*} H \, dt = H \, (t_* - t) = -Ht, \qquad (5.30)$$

in which, without loss of generality, we have set $t_* = 0$ so $N > 0$. With this definition of time, the solution for ϕ, expressed in terms of N, is

$$\phi(N) = \lambda + \mu e^{3N} = \lambda + (\phi_* - \lambda) e^{3N}. \qquad (5.31)$$

From this solution we obtains

$$\dot{\phi}(N) = -3H\mu e^{3N} = 3H(\lambda - \phi_*) e^{3N}. \qquad (5.32)$$

Equations (5.31) and (5.32) indicate that different trajectories in the phase space $(\phi, \dot{\phi})$ are parametrized by λ with N being the parameter along each trajectory. That is,

$$\phi = \phi(N, \lambda), \quad \dot{\phi} = \dot{\phi}(N, \lambda). \qquad (5.33)$$

In other words, the variables (N, λ) may be regarded as another set of coordinates in the phase space. Therefore one can perform a simple coordinate transformation in phase space and obtain N and λ as functions of $(\phi, \dot{\phi})$, yielding

$$N = N\left(\phi, \dot{\phi}\right) = \frac{1}{3} \ln \left(\frac{\dot{\phi}}{\dot{\phi} + 3H(\phi - \phi_*)} \right), \qquad (5.34)$$

$$\lambda = \lambda\left(\phi, \dot{\phi}\right) = \phi + \frac{\dot{\phi}}{3H}. \qquad (5.35)$$

In the present case at hand with $c_s = 1$, Eq. (5.34) by itself is enough to find the δN formula. However, in order for our simple analysis here to be helpful for the more complicated case when $c_s \neq 1$, we follow an intermediate step for the derivation of the δN formula as follows.

Instead of $(\phi, \dot{\phi})$, we may also use the coordinate system (ϕ, λ) in the phase space. One advantage of this choice is that the coordinate λ is a constant of integration along each trajectory. Therefore its perturbation $\delta\lambda$ can be evaluated at any point along the trajectory. With this choice, we can express N as $N = N(\phi, \lambda)$. More specifically, from Eq. (5.31) we obtain $N = N(\phi, \lambda)$ as

$$N = \frac{1}{3} \ln \left(\frac{\phi - \lambda}{\phi_* - \lambda} \right). \tag{5.36}$$

Now, we are ready to calculate δN by setting $\phi \to \phi + \delta\phi$ and $\lambda \to \lambda + \delta\lambda$. Up to the second order in perturbations we get

$$\delta N = \frac{\partial N}{\partial \phi} \delta\phi + \frac{\partial N}{\partial \lambda} \delta\lambda + \frac{1}{2} \frac{\partial^2 N}{\partial \phi^2} \delta\phi^2 + \frac{1}{2} \frac{\partial^2 N}{\partial \lambda^2} \delta\lambda^2 + \frac{\partial^2 N}{\partial \phi \partial \lambda} \delta\phi\delta\lambda. \tag{5.37}$$

Now we identify the perturbations $\delta\phi$ and $\delta\lambda$ with those evaluated on the flat hypersurface at or after the mode of interest has left the horizon. For $\delta\phi$, this is simple. However, for $\delta\lambda$, we need its relation with $\delta\phi$ and $\delta\dot{\phi}$. This is easily obtained from Eq. (5.35) yielding

$$\delta\lambda = \delta\phi + \frac{\delta\dot{\phi}}{3H}. \tag{5.38}$$

After the system has reached the attractor limit, the dominant solution for $\mathcal{R}_c$ is the constant solution (the term containing C_1 in Eq. (5.22)). As a result, on superhorizon scales $\delta\dot{\phi} = 0$ and hence $\delta\lambda = \delta\phi$. Inserting this into Eq. (5.37), we obtain our final δN formula

$$\delta N = \frac{\delta\phi}{3(\phi_* - \lambda)} + \frac{\delta\phi^2}{6(\phi_* - \lambda)^2}. \tag{5.39}$$

Note that the above expression for δN is calculated at the end of non-attractor phase $\phi = \phi_*$.

As mentioned above, we could also obtain δN formula directly from Eq. (5.34). Perturbing Eq. (5.34) till second order in fields and noting that $\delta\dot\phi = 0$ on superhorizon scales, we get

$$
\begin{aligned}
\delta N &= \frac{\partial N}{\partial \phi}\delta\phi + \frac{1}{2}\frac{\partial^2 N}{\partial \phi^2}\delta\phi^2 \\
&= -\frac{H\delta\phi}{\dot\phi + 3H(\phi - \phi_*)} + \frac{3H^2\delta\phi^2}{2(\dot\phi + 3H(\phi - \phi_*))^2}
\end{aligned}
\tag{5.40}
$$

Using Eq. (5.35) to eliminate $\dot\phi$ in favor of λ, this also reduces to Eq. (5.39) when calculated at $\phi = \phi_*$.

Having obtained δN to second order in perturbations, either in the form of Eq. (5.39) or Eq. (5.40), we can obtain the amplitude of the local-type non-Gaussianity f_{NL}^{local} which is defined via

$$
\mathcal{R}_c = \mathcal{R}_{c,g} + \frac{3}{5}f_{NL}\mathcal{R}_{c,g}^2,
\tag{5.41}
$$

where $\mathcal{R}_{c,g}$ is the Gaussian part of curvature perturbations. Comparing this with Eq. (5.39) or Eq. (5.40) we find

$$
f_{NL}^{\text{local}} = \frac{5}{2}.
\tag{5.42}
$$

This is very interesting. We have obtained an observable value for f_{NL}^{local} even with near scale-invariant power spectrum. Therefore, Maldacena's consistency condition is violated. As mentioned before, one key assumption in proving Maldacena's consistency condition is that $\mathcal{R}_c$ is frozen on superhorizon scales. However, in our non-attractor model, with $\eta \simeq -6$ and $\nu \simeq -3/2$, we conclude from Eq. (5.23) that $\mathcal{R}_c \propto (k\tau)^{2\nu} \propto (-k\tau)^{-3}$. As a result, for superhorizon modes $\mathcal{R}_c$ grows exponentially. This clearly violates the necessary condition for the validity of Maldacena's consistency condition. Having this said, note that we cannot allow $\mathcal{R}_c$ to grow indefinitely, otherwise the system becomes non-perturbative. This is avoided by terminating the non-attractor phase at the time $t = t_*$ followed by the second attractor phase in which $\mathcal{R}_c$ is conserved on super-horizon scales as usual.

The δN analysis to calculate f_{NL}^{local} for the general value of c_s is somewhat complicated and we defer it to Appendix A. Here we only present the final result

$$f_{NL}^{\text{local}} = \frac{5}{4c_s^2} \left(1 + c_s^2\right). \qquad (5.43)$$

This result is valid for any values of c_s as we have not assumed $c_s \ll 1$ in our analysis. In addition, when $c_s = 1$ we obtain $f_{NL}^{\text{local}} = 5/2$ which agrees with Eq. (5.42). Of course, there are strong observational constraints on f_{NL}^{local} from Planck satellite data, so in order to be consistent with these data one cannot take c_s very small. As a result one requires $c_s \lesssim 1$ to obtain $f_{NL}^{\text{local}} = \text{few}$ in order to be consistent with the observations.

We also mention that the bispectrum analysis for this model using the field-theoretical in-in formalism was performed in Ref. [65]. Interestingly, our δN result for f_{NL}^{local} agrees exactly with the result obtained from the in-in formalism.

To summarize, in this chapter we have presented another nontrivial application of δN formalism for the non-attractor backgrounds. As we saw, the key feature of the non-attractor background is that $\mathcal{R}_c$ is not frozen on superhorizon scales. For example, for the model studied here $\mathcal{R}_c \propto a^3$ on superhorizon scales. As a result, the background trajectory is a function of both the field value ϕ and its derivative $\dot{\phi}$. This is a very interesting example which reminds us that in general δN should be expressed in the two-dimensional phase space of $(\phi, \dot{\phi})$. This is in contrast to the usual slow-roll models in which $\dot{\phi}$ is uniquely determined in terms of ϕ so one can assume $N = N(\phi)$.

Since the scalar field trajectory is determined by two parameters, one can specify the scalar field value and its derivative at some initial epoch, ϕ_i and $\dot{\phi}_i$ [63]. However, in this chapter we find it more convenient to use the coordinate (ϕ, λ) in the phase space. We use this methodology to obtain $f_{NL}^{\text{local}} = 5(1 + c_s^2)/4c_s^2$, valid for all value of c_s.

Another important implications of the non-attractor model is that Maldacena's consistency condition for the single-field inflation is violated. Again, this is because $\mathcal{R}_c$ is not frozen on superhorizon scales so one cannot simply absorb the effects of *conserved* $\mathcal{R}_c$ by re-scaling the effective scale factor via $a \to a\,e^{\mathcal{R}_c}$ as usually done to prove Maldacena's consistency condition [67]. We mention that so far the non-attractor models are the only self-consistent, single-field inflation scenarios based on a Bunch–Davies initial state, that generate a scale-invariant power spectrum and an observable squeezed-limit non-Gaussianity which violate Maldacena's consistency condition. For related works in this direction see Ref. [75–78].

As briefly mentioned above, our result for f_{NL}^{local} agrees exactly with the result obtained from the in-in formalism. However, this should not come as a surprise. The reason is that in our model the dynamics responsible for generating local-type non-Gaussianities are on superhorizon scales. In addition, there are no intrinsic non-Gaussianities generated around the time of horizon crossing so $\delta\phi$ has a trivial statistics at the time of horizon crossing. Therefore, the δN method, relying on the classical background field evolution with the known statistics of the initial field perturbations, is expected to be applicable and should yield the same value for f_{NL}^{local} as obtained from in-in formalism.

Application of δN formalism:
Inflation with local features

6.1 Motivation

In this chapter we present another non-trivial example of δN formalism in models of inflation in which there are local features in primordial power spectrum and bispectrum. Before presenting the detailed δN application, let us first present the motivations for considering these effects.

As we learned so far, the simplest models of inflation predict almost scale-invariant, almost Gaussian and almost adiabatic perturbations on CMB which are in good agreements with observations. Having said this, there are hints of deviation from this simple picture on large scales, say $\ell \lesssim 50$, such as the shortage of power, hemispherical dipole asymmetry and quadrupole–octopole alignment [39, 79, 80]. These anomalies may simply be due to cosmic variance. Alternatively, these effects can be associated with non-trivial dynamics during early stage of inflation which deviates from the above simple picture. In particular, one may try to construct a model of inflation in which there are non-trivial dynamics happening when the large scale perturbations exit the horizon. On the physical grounds, one expects that only these large scale modes (say $\ell \lesssim 50$) are affected while small scale perturbations are unaffected.

There have been many attempts in the literature to generalize local features during inflation which may have non-trivial predictions on the power spectrum and bispectrum, see e.g. Refs. [81–83] for early works in this direction. As mentioned above, this is partly motivated from glitches in the CMB angular power spectrum at $\ell \lesssim 50$. These local features may originate from models of high energy physics, via particle creation or through field annihilation during inflation etc. Usually in these models local features are generated by temporal violations of slow-roll conditions during inflation in an *ad hoc* manner with no dynamical realization as to how to violate the slow-roll conditions. Here we provide a consistent dynamical mechanism to generate local features in the power spectrum and bispectrum as presented in Ref. [84].

The model consists of a simple chaotic inflation potential, $m^2\phi^2/2$, which is coupled to a heavy field χ. After ϕ reaches a critical value $\phi = \phi_c$ ($\gg M_P$), χ becomes tachyonic and a rapid waterfall phase transition takes place. In this view the model is similar to hybrid inflation [61, 62]. However, the important difference is that the potential is not vacuum dominated and the chaotic-type inflation potential is the main driving source of inflation. Furthermore, inflation will continue even after the waterfall phase transition. In this scenario, the entire period of inflation may be divided into three stages. During the first stage $\phi > \phi_c$, inflation proceeds similar to chaotic inflation. The second stage, at which $\phi \lesssim \phi_c$, is short and the heavy field χ becomes tachyonic and a waterfall phase transition takes place such that χ settles to its global minimum. As a result, this effect causes a small change in the inflaton effective mass and $m^2 \to m_+^2 = m^2(1+C)$ in which $C \ll 1$. The final stage of inflation proceeds as in chaotic inflation. In order to bring the local feature into the CMB observable window, we require that the waterfall phase transition takes place about 55–60 e-folds before the end of inflation (i.e. the feature happens in the first 5 e-folds or so of inflation).

The crucial ingredients of our analysis are the dynamics of the waterfall quantum fluctuations and their contributions to the power spectrum of curvature perturbations. Our approach in studying the dynamics of waterfall fluctuations is similar to Refs. [85–87]. Since we assume the waterfall is sharp, the waterfall contribution to $\mathcal{R}_c$ is narrowly localized around the modes which leave the horizon at the time of waterfall instability. In addition, because of the intrinsic non-Gaussian nature of the waterfall contribution, which is in the form of $\mathcal{R}_c \sim \delta\chi^2$, a large spiky non-Gaussianity is generated. We stress that this is a local dynamical effect intrinsic to the waterfall quantum fluctuations which does not show up in other models studying local feature in *ad hoc* manner. In this view these spiky effects in primordial correlations are observationally testable as a genuine effect of our model.

The δN formalism in this chapter is employed to calculate the power spectrum and the bispectrum. Because of the $\delta\chi^2$ nature of the waterfall quantum fluctuations, the non-Gaussianity analysis is somewhat technical. For a first reading, the reader may skip the non-Gaussianity analysis. In addition, we have relegated some technical details about the power spectrum of the waterfall quantum fluctuations and higher order δN perturbations into appendices.

6.2   The model

In this section we present our model for generating local features in curvature perturbation during inflation. We consider the simplest inflationary scenario, the chaotic potential $m^2\phi^2/2$, in which we assume the mass of inflaton field undergoes a sudden dynamical change at $\phi = \phi_c$. Our goal is to see how a sudden small change in the inflaton mass can be modeled in a consistent dynamical manner and then look for its observational consequences in correlation functions of primordial perturbations. We employ the idea of waterfall phase transition to generate local feature during inflation. The setup we consider is the following

$$V = \frac{m^2}{2}\phi^2 + \frac{\lambda}{4}\left(\chi^2 - \frac{M^2}{\lambda}\right)^2 + \frac{g^2}{2}\phi^2\chi^2. \tag{6.1}$$

Here ϕ is the inflaton field and χ is the waterfall field. Formally the above potential is identical to the model of hybrid inflation [61, 62]. However, as we stressed before, in our setup inflation is mainly driven by the field ϕ so it effectively proceeds as in chaotic model with potential $V \simeq m_{eff}^2 \phi^2/2$ in which the effective mass m_{eff} undergoes a small but rapid change at $\phi = \phi_c$. The role of the heavy waterfall field χ is to induce this change in mass.

Like in chaotic inflation, inflation starts at $\phi = \phi_i \gg M_P$ so one can obtain about 60 *e*-folds of inflation to solve the horizon and flatness problems. We assume that the waterfall field is very heavy during the first stage of inflation so it is locked at its instantaneous minimum $\chi = 0$. However, once the inflaton field reaches the critical value $\phi = \phi_c \equiv M/g$, the waterfall field χ becomes tachyonic and it quickly rolls towards its global minimum $\chi_{min}^2 = M^2/\lambda$. We work in the limit that this waterfall phase transition is sharp, i.e. the waterfall field settles to its global minimum in less than an *e*-fold. The final stage of inflation after waterfall transition for $\phi > \phi_c$ continues as in chaotic inflation model but now the effective mass of inflaton, m_+, is

$$m_+^2 = m^2 + g^2 \langle \chi^2 \rangle = m^2 \left(1 + C \right), \qquad (6.2)$$

in which the parameter C is defined by

$$C \equiv \frac{g^2 M^2}{\lambda m^2}. \qquad (6.3)$$

We are interested in the case in which the change in inflaton mass is small so $C \ll 1$. In addition, in order to bring this local feature into the CMB observable window, we assume that the waterfall stage begins at about 55 *e*-folds before the end of inflation. To a good approximation, $V \simeq m^2 \phi^2/2 + M^4/4\lambda$ so $V(\phi_c) = m^2 \phi_c^2/2(1 + C/2)$. Demanding $C \ll 1$ corresponds to the assumption that the inflationary potential is dominated by the $m^2 \phi^2/2$ term. This is in contrast to the standard hybrid inflation scenario in which the potential is dominated by the vacuum and $C \gg 1$.

It is more convenient to work with the number of *e*-folds N as the new clock in which N is related to cosmic time t via $dN = Hdt$. Denoting the time when the waterfall phase transition starts by N_c, we further define $n \equiv N - N_c$. Hence $n < 0$ for the period before the phase transition while for the period after the transition $n > 0$. The end of waterfall transition when χ has settled down to its global minimum is denoted by $n = n_f$. With these definitions, inflation in our model has the three stages: (a) $n < 0$, (b) $0 \leq n \leq n_f$, and (c) $n_f < n < N_e - N_c$ where N_e represents the time when inflation ends. As mentioned, we set $N_e - N_c \sim 55$ so that the waterfall transition falls into the observable window. We are interested in a sharp waterfall phase transition corresponding to $n_f \lesssim 1$. The key ingredient in our analysis is the dynamics of the waterfall quantum fluctuations during this short stage and their contribution to curvature perturbations. This is where the δN formalism comes into play in a non-trivial way.

For convenience we introduce the dimensionless parameters α and β, related to mass parameters m and M, via

$$\alpha \equiv \frac{m^2}{H^2} \simeq \frac{6M_P^2}{\phi^2} \simeq \frac{6g^2 M_P^2}{M^2}, \quad \beta \equiv \frac{M^2}{H^2} \simeq \frac{6M^2 M_P^2}{m^2 \phi^2} \simeq \frac{6g^2 M_P^2}{m^2}, \quad (6.4)$$

where the last approximate equalities are valid assuming $\phi \simeq \phi_c$. The validity of slow-roll assumption during the first and third inflationary stages requires $\alpha \ll 1$. Furthermore, demanding that the waterfall field is heavy requires that $\beta \gg 1$ and therefore $g^2 \gg m^2/M_P^2$. In addition, from the COBE normalization we have $m/M_P \sim 10^{-6}$ and consequently $g^2 \gg 10^{-12}$. Furthermore, as mentioned before, we assume that the waterfall phase transition occurs about 55 e-folds before the end of inflation so it falls within the CMB observational window. In order for inflation to continue effectively in the form of standard chaotic inflation for long enough period after the waterfall phase transition we require $\phi_c \gtrsim 10 M_P$ so $g^2 \lesssim 10^{-2} M^2/M_P^2$. Combining this requirement with $g^2 \gg m^2/M_P^2$ we obtain $m^2 \ll 10^{-2} M^2$ or $M \gg 10^{-5} M_P$. Finally, from the form of parameter C we conclude that $g^2/\lambda \ll 10^{-2} C$. As an example if we assume $C \sim 10^{-2}$, then we get $g^2/\lambda \ll 10^{-4}$ and $\lambda = C^{-1} g^2 M^2/m^2 \sim 10^2 g^2 M^2/m^2 \gg 10^4 g^2 \gg 10^{-8}$.

6.2.1 *Dynamics of inflaton*

Here we study the dynamics of inflaton field (see Fig. 6.1). As mentioned before, during the first and second stages, inflation proceeds somewhat similar to standard chaotic inflation with the potential given by

$$V^-(\phi) = \frac{1}{2} m^2 \phi^2 + \frac{M^4}{4\lambda}. \tag{6.5}$$

During the short second stage in which the waterfall field χ becomes tachyonic it grows until the self-interaction term $\lambda \chi^4$ becomes important. Then the self-interaction induces a large effective mass and χ settles down to its global minimum. We represent this time by n_f. After that, the waterfall field rolls across the valley determined by $\partial_\chi V(\phi, \chi) = 0$, yielding

$$\chi^2 = \chi^2_{min} \equiv \frac{M^2}{\lambda} - \frac{g^2}{\lambda} \phi^2. \tag{6.6}$$

As we see from the equation above the local value of χ during the third stage is dictated solely by the value of the inflaton field. Consequently, the inflaton field during the third stage experiences an effective potential given by

$$V^+_{eff}(\phi) = V(\phi, \chi(\phi)) = \frac{1}{2} m^2 (1 + C) \phi^2 - \frac{g^4}{4\lambda} \phi^4. \tag{6.7}$$

Solving the equations of motion for ϕ for the first and second stages with the slow-roll conditions we obtain

$$-4M_P^2 (N - N_c) = -4M_P^2 n = \phi(n)^2 - \phi_c^2 \left[1 - C \ln \left(\frac{\phi}{\phi_c} \right) \right], \tag{6.8}$$

while for the third stage we have

$$8M_P^2 (N_e - N) = -\phi_e^2 + \phi(N)^2 + \frac{1+C}{C} \phi_c^2 \ln \left[\frac{(1+C)/C - (\phi_e/\phi_c)^2}{(1+C)/C - (\phi/\phi_c)^2} \right]$$

$$= 2\phi^2(N) - 2\phi_e^2 + \frac{C}{2} \frac{\phi^4(N) - \phi_e^4}{\phi_c^2} + \mathcal{O}(C^2). \tag{6.9}$$

As in chaotic models, inflation ends at $\phi = \phi_e$ when either of the two slow-roll conditions $\epsilon, \eta \ll 1$ is violated corresponding to $\phi_e = \sqrt{2} M_P$.

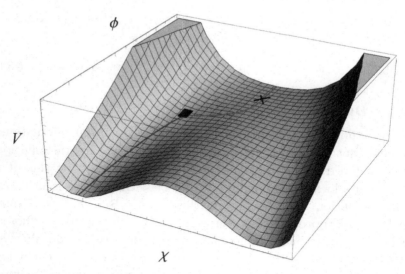

ϕ

V

χ

Fig 6.1   The evolution of inflaton and waterfall fields (smoothened over a horizon-size patch). The parameters used to produce the plot are exaggerated so that the different stages of inflation could be distinguished. The upper solid blue curve represents the first stage of inflation before the transition starts at $\phi = \phi_c$. In this stage $(\phi < \phi_c)$ the waterfall field is trapped at zero value $\chi = 0$ while the inflaton field rolls as in single-field inflation models. The middle solid red curve shows the evolution of the fields in the transition region, say $\phi > \phi_c$ and $n < n_f$, where the waterfall field becomes unstable and hence it bends the trajectory towards the valley of local minima. The lower solid purple curve represents the trajectory of fields along the valley of local minima for $n > n_f$. As stressed, after the completion of the transition, the dynamics of inflaton field would be "effectively" as in single-field models with a modified mass.

Note that in this chapter the slow-roll parameters are defined via derivatives of potential as

$$\epsilon \equiv \frac{M_P^2}{2}\left(\frac{V_\phi}{V}\right)^2, \quad \eta \equiv M_P^2 \frac{V_{\phi\phi}}{V}, \tag{6.10}$$

and around the epoch of the phase transition, they are approximately expressed in terms of the parameter α as

$$\epsilon \simeq \eta \simeq \frac{\alpha}{3}. \qquad (6.11)$$

6.2.2   Dynamics of waterfall field

Now we study the dynamics of the waterfall quantum fluctuations. An important point in the current setup, similar to hybrid inflation, is that before the waterfall transition χ is very heavy so that it is locked at its local minimum $\chi = 0$ and classically there is no background evolution for χ. Following the logic of Ref. [85] we assume that for some appropriate smoothing size, $\delta\chi^2$ is treated as a "homogeneous classical" background quantity which varies smoothly over scales larger than the Hubble scale. To set the stage let us review two important points about the dynamics of the inhomogeneous perturbations of the waterfall field. First, in the following we show that due to large tachyonic mass of the waterfall field, the long mode fluctuations of χ_k become unstable after some transition point $n_t(k)$, after which they grow exponentially. In the growing phase, the χ_k mode becomes a real field and henceforth can be regarded as a classical field [88, 89]. It should be noticed that as a result of large tachyonic mass, most of the χ_k fluctuations become classical perturbation well before exiting the horizon. Secondly, it can be seen that shortly after the time that χ_k becomes unstable, the gradient term $\nabla\chi_k$ can be ignored compared to time derivative term on every smoothing scale larger than the inverse of (the absolute value of) the waterfall tachyonic mass. Specifically, this shows that for smoothened patches of the horizon scale, the χ field evolves locally, justifying the applicability of the separate universe approach in the current setup.

Consequently, for sufficiently large scales, say scales larger than the comoving scale of the present Hubble horizon size, one can calculate the mean value $\langle \delta\chi^2(n) \rangle$ and assign its fluctuation via

$$\Delta\chi^2(n,\mathbf{x}) \equiv \delta\chi^2(n,\mathbf{x}) - \langle \delta\chi^2(n) \rangle. \qquad (6.12)$$

In this view $\langle \delta\chi^2(n) \rangle$ represents the homogeneous background value while $\Delta\chi^2(n,\mathbf{x})$ seeds the curvature perturbations on superhorizon scales [85, 89]. With this prescription, we study the dynamics of the background waterfall field and its quantum fluctuations. It should be noted that the time dependence of H is ignored in the following calculations which is indeed a good approximation during inflation. The dynamics of the background waterfall field is given by

$$\chi'' + 3\chi' + \left(-\beta + g^2\frac{\phi^2}{H^2} + 3\lambda\frac{\chi^2}{H^2}\right)\chi = 0, \qquad (6.13)$$

where here and below the prime denotes a derivative with respect to clock n with $dn = Hdt$. Neglecting the self-interaction term $\lambda\frac{\chi^2}{H^2}$ during the second stage when the transition is still in progress, we obtain the following approximate solution for the background waterfall field [85]

$$\chi(n) \simeq \chi(n=0)\exp\left[\frac{2}{3}\epsilon_\chi n^{3/2}\right], \qquad (6.14)$$

where

$$\epsilon_\chi \simeq \sqrt{\frac{2}{3}\alpha\beta}. \qquad (6.15)$$

It is worth looking at the relations among the model parameters. The model has four parameters: M, m, g and λ. However, we have introduced six non-dimensional parameters: α, β, ϵ, η, ϵ_χ and C. Some of these parameters are time dependent, but at leading order in the slow-roll approximation, we can consider them as constants. Therefore, as another convenient set of independent parameters, we can choose ϵ, C, ϵ_χ and λ as free parameters yielding

$$\alpha = 3\epsilon, \quad \beta = \frac{\epsilon_\chi^2}{2\epsilon}, \quad \eta = \epsilon, \quad g^2 = \frac{6\lambda\epsilon^2 C}{\epsilon_\chi^2},$$
$$M^2 \simeq \frac{12\lambda\epsilon C}{\epsilon_\chi^2}M_P^2, \quad m^2 \simeq \frac{72\lambda\epsilon^3 C}{\epsilon_\chi^4}M_P^2. \qquad (6.16)$$

Going to Fourier space and neglecting the self-interaction term, the dynamics of the waterfall field quantum fluctuations is governed by

$$\delta\chi_{\mathbf{k}}'' + 3\delta\chi_{\mathbf{k}}' + \left(\frac{k^2}{a^2 H^2} - \beta + g^2\frac{\phi^2}{H^2}\right)\delta\chi_{\mathbf{k}} = 0, \qquad (6.17)$$

in which

$$\delta\chi_{\mathbf{k}} = \int \frac{d^3x}{(2\pi)^{3/2}} \delta\chi(\mathbf{x}) e^{-i\mathbf{k}\cdot\mathbf{x}} = a_{\mathbf{k}}\chi_k(n) + a^{\dagger}_{-\mathbf{k}}\chi^*_k(n). \qquad (6.18)$$

Here $a_{\mathbf{k}}$ and $a^{\dagger}_{\mathbf{k}}$ respectively are the annihilation and creation operators associated with the waterfall quantum fluctuations defined with respect to the Minkowski vacuum and $\chi_k(n)$ represents the positive frequency wave function. Plugging the background solution of ϕ from Eq. (6.8) into equation of $\delta\chi_{\mathbf{k}}$, Eq. (6.17), yields

$$\delta\chi''_{\mathbf{k}} + 3\delta\chi'_{\mathbf{k}} + \left(\frac{k^2}{k_c^2}e^{-2n} - \epsilon_\chi^2 n\right)\delta\chi_{\mathbf{k}} = 0, \qquad (6.19)$$

in which k_c represents the comoving momentum associated to the mode which exits the Hubble radius at the time of waterfall phase transition defined via $k_c = Ha(n=0)$. As we see from Eq. (6.19), $\epsilon_\chi H^2$ is a measure of the effective tachyonic mass of the waterfall field when the tachyonic instability is in progress. The assumption that the waterfall phase transition is sharp requires that $\epsilon_\chi \gg 1$. The normalization of $\chi_{\mathbf{k}}$ is fixed by the canonical commutation relation, yielding

$$\delta\chi_{\mathbf{k}}\delta\chi'^{*}_{\mathbf{k}} - \delta\chi^*_{\mathbf{k}}\delta\chi'_{\mathbf{k}} = \frac{iH^2}{k_c^3 e^{3n}}. \qquad (6.20)$$

For large and negative values of n, Eq. (6.19) can be solved using the WKB approximation. Taking into account the normalization condition (6.20) and choosing the standard Minkowski positive frequency for the initial conditions in the limit $n \to -\infty$, the result to the first order in the WKB approximation is obtained to be [87]

$$\delta\chi_k(n) = \frac{H}{\sqrt{2k_c^3}} \frac{e^{-3n/2}}{((k/k_c)^2 e^{-2n} - \epsilon_\chi^2 n)^{1/4}}$$
$$\times \exp\left[-i\int^n \left((k/k_c)^2 e^{-2n'} - \epsilon_\chi^2 n'\right)^{1/2} dn'\right]. \qquad (6.21)$$

After waterfall phase transition, $n > 0$, some (low k) modes of waterfall field fluctuations become tachyonic and behave like classical random fields [86]. In particular, there are modes which become tachyonic even before leaving the horizon. In order to find the dynamics of the waterfall quantum fluctuations, we divide the modes into large and small modes, denoted below by the subscripts L and S respectively. Large modes are those which leave the horizon before the onset of waterfall phase transition at $n = 0$ while small modes are those which are subhorizon at the time $n = 0$.

The large modes leave the horizon before the time of waterfall transition while after horizon crossing their profile falls off like $\propto e^{-3n/2}$ as can be seen from Eq. (6.21). At the time of waterfall transition, $n = 0$, the WKB approximation fails but apart from a factor of order unity the amplitude of the wave function at $n = 0$ is approximately given by

$$|\delta\chi_{\mathbf{k}}^{L}(n = 0)| \simeq \frac{H}{\sqrt{2\epsilon_\chi k_c^3}}. \tag{6.22}$$

After the transition, $n > 0$, these modes obey the same equation as the classical trajectory, so using Eq. (6.14), one obtains

$$|\delta\chi_{\mathbf{k}}^{L}(n > 0)| \simeq \frac{H}{\sqrt{2\epsilon_\chi k_c^3}} \exp\left(\frac{2}{3}\epsilon_\chi n^{3/2}\right). \tag{6.23}$$

The situation for the small scales is somewhat non-trivial and requires careful considerations. As mentioned above the effective tachyonic mass of $\delta\chi_{\mathbf{k}}$ is at the order of $\epsilon_\chi H \gg H$. As a result some modes become tachyonic even before crossing the Hubble radius. For modes which become tachyonic the spatial gradient term is negligible compared to the tachyonic mass and the evolution of $\delta\chi_{\mathbf{k}}$ becomes identical to that of the background field. Therefore, for each mode it is important to determine the time when it becomes tachyonic. Representing the time when the mode k becomes tachyonic by $n_t(k)$, from Eq. (6.19) one obtains

$$n_t(k)\,e^{2n_t(k)} = \left(\frac{k}{\epsilon_\chi k_c}\right)^2. \tag{6.24}$$

In the following we shall call this time the time of "classicalization". The solution of the above algebraic equation are given in terms of the

Lambert W function $W(z)$,

$$n_t(k) = \frac{1}{2}W(z); \quad z = 2\left(\frac{k}{\epsilon_\chi k_c}\right)^2. \tag{6.25}$$

However, we only need an approximate solution for $n_t(k)$ because $n_t(k) \lesssim 1$ for the parameters corresponding to a sharp phase transition.

Using the WKB solution Eq. (6.21) until $n = n_t(k)$ we obtain

$$\delta\chi_{\mathbf{k}}^S(n) = \frac{H}{\sqrt{2k\,k_c}}e^{-n}; \quad n < n_t(k). \tag{6.26}$$

As mentioned before, after $n = n_t(k)$ this mode evolves like the classical background solution, Eq. (6.14). Matching the solution with the classical trajectory at $n = n_t(k)$, for $n > n_t(k)$ we approximately obtain

$$\delta\chi_{\mathbf{k}}^S(n) = \frac{H}{\sqrt{2k\,k_c}}e^{-n_t}\exp\left[\frac{2}{3}\epsilon_\chi\left(n^{3/2} - n_t^{3/2}\right)\right]; \quad n > n_t(k). \tag{6.27}$$

It is convenient to project $\delta\chi_k^S(n)$ at the time of waterfall transition $n = 0$ as if all the modes were tachyonic at $n > 0$, as viewed in Refs. [85, 87]. This yields

$$\delta\chi_{\mathbf{k}}^S(0) = \frac{H}{\sqrt{2k\,k_c}}\exp\left[-n_t(k) - \frac{2}{3}\epsilon_\chi n_t^{3/2}(k)\right]. \tag{6.28}$$

Because of the exponential growth of the waterfall field fluctuations after the waterfall transition, the expectation value $\langle\delta\chi^2\rangle$ is not negligible and its rms value soon starts to behave as a classical field $\chi = \sqrt{\langle\delta\chi^2\rangle}$ [85]. This implies that $\langle\delta\chi^2\rangle$ can be viewed as a classical background for an observer within each Hubble horizon region [85–87]. Therefore we need to calculate $\langle\delta\chi^2\rangle$ which in turns is given in terms of its power spectrum $\mathcal{P}_{\delta\chi}$ via

$$\langle\delta\chi^2\rangle = \int\frac{d^3k}{(2\pi)^3}|\delta\chi_{\mathbf{k}}|^2 = \int\frac{d^3k}{(2\pi)^3}P_\chi(k) = \int\frac{dk}{k}\mathcal{P}_{\delta\chi}(k), \tag{6.29}$$

where the power spectrum is defined as usual by

$$\langle \delta\chi_{\mathbf{k}}\delta\chi_{\mathbf{q}} \rangle \equiv (2\pi)^3 \delta^3 (\mathbf{k}+\mathbf{q}) P_\chi (k) , \quad \mathcal{P}_{\delta\chi} \equiv \frac{k^3}{2\pi^2} P_\chi (k) . \qquad (6.30)$$

Using Eqs. (6.23) and (6.27) the power spectrum of the waterfall quantum fluctuations at $n = 0$ can be read off as

$$\mathcal{P}_{\delta\chi}(k;0) = \begin{cases} \dfrac{H^2}{4\pi^2 \epsilon_\chi} \left(\dfrac{k}{k_c}\right)^3 ; & k < k_c, \\[3mm] \dfrac{H^2 \epsilon_\chi^2}{4\pi^2} n_t(k) \exp\left[-\dfrac{4}{3}\epsilon_\chi n_t^{3/2}(k)\right] ; & k > k_c. \end{cases} \qquad (6.31)$$

The details of the analysis to calculate $\langle \delta\chi^2 \rangle$ are presented in Appendix B where it is shown that $\langle \delta\chi^2 \rangle$ is dominated by the small scale modes and

$$\langle \delta\chi^2(0) \rangle \simeq \langle \delta\chi^2(0) \rangle_S \simeq \frac{3\epsilon_\chi^{4/3} H^2}{16\pi^2} . \qquad (6.32)$$

From Eq. (6.31) we see that there is a sharp peak in the spectrum of waterfall fluctuations. In order to estimate the width of the peak, we expand the above spectrum near the location of its peak. Solving $\partial \mathcal{P}_{\delta\chi}/\partial n_t = 0$, the position of peak is found to be

$$n_t(k_{max}) = \left(\frac{1}{2\epsilon_\chi}\right)^{2/3} . \qquad (6.33)$$

Expanding the spectrum around k_{max} yields

$$\mathcal{P}_{\delta\chi}(k;0) \simeq \frac{H^2 \epsilon_\chi^2}{4\pi^2} \left(\frac{1}{2e\epsilon_\chi}\right)^{2/3} \exp\left[-\frac{(n_t(k)-n_t(k_{max}))^2}{2\sigma_{n_t}^2}\right] , \qquad (6.34)$$

in which

$$\sigma_{n_t} = \sqrt{\frac{2}{3}} \left(\frac{1}{2\epsilon_\chi}\right)^{2/3} = \sqrt{\frac{2}{3}} n_t(k_{max}) . \qquad (6.35)$$

However, we are interested in the width of spectrum in the momentum space, $\sigma_*(k)$. Using Eq. (6.24) one obtains

$$\sigma_*(k) = \left(1 + \frac{1}{2n_t(k)}\right)\sigma_{n_t} k_{max} \qquad (6.36)$$

$$\simeq \left(\sqrt{\frac{1}{6}} + \mathcal{O}\left(\epsilon_\chi^{-2/3}\right)\right) k_{max} \simeq 0.4 k_{max}. \qquad (6.37)$$

This shows that for large enough ϵ_χ the width of the waterfall power spectrum is independent of the sharpness of the phase transition which can also be verified numerically.

6.3 δN formalism in models with localized feature

Having presented the dynamics of field equation, we are now ready to present the δN formalism to calculate the curvature perturbations. In order to do this properly we trace back the number of e-folds from the end of inflation until the time of horizon crossing for each given mode of interest. To avoid confusion we represent the number of e-folds counted *backward* in time from the end of inflation by $\mathcal{N}$, so $\mathcal{N} \equiv N_e - N$. To employ δN formalism we have to express $\mathcal{N}$ in terms of the background fields $\phi(n)$ and $\chi^2(n)$, where the latter is smoothened on each Hubble patch.

For the modes which exit the horizon after the waterfall transition, one easily finds the curvature perturbation on comoving slices from Eq. (6.9) yielding

$$\mathcal{R}_c = \delta \mathcal{N} = \left(1 + \frac{C}{2}\frac{\phi^2}{\phi_c^2}\right)\frac{\phi\delta\phi}{2M_P^2}, \qquad (6.38)$$

where, as usual, $\delta\phi$ is evaluated on flat hypersurfaces. Note that compared to simple chaotic model, we have the additional term containing the small parameter C which represents the effects of change in inflaton mass from m to m_+ as given in Eq. (6.2).

However, calculating the curvature perturbation for the modes which exit the horizon before and during the waterfall transition needs careful considerations. For these modes we follow the same procedures as employed in Ref. [86]. We note that the duration of the waterfall stage depends on the classical value of the waterfall field on every smoothing patch. As mentioned previously, because of the large tachyonic mass of the waterfall field there exists modes whose spatial gradient can be neglected and which behave classically even before crossing the horizon and hence affect the classical trajectory. Noting that the relation $\mathcal{R}_c = \delta\mathcal{N}$ is valid on scales over which small scale inhomogeneities can be smoothened out with a negligible influence on the geometry, we take the smoothing scale to be slightly larger than the comoving scale associated with the wavelength of the last mode which becomes tachyonic.

As discussed before, during the third inflationary stage (after the completion of the waterfall phase transition) inflation proceeds as in chaotic inflation with a slight change in the effective mass of inflaton. Consequently the end of inflation is determined uniquely by the value of ϕ as

$$\phi = \phi_e \approx \sqrt{2}\, M_P. \tag{6.39}$$

From this point till the end of waterfall transition, $\mathcal{N}$ is obtained by Eq. (6.9) so

$$\mathcal{N} = \frac{1}{4M_P^2}\left[\phi^2 - \phi_e^2 + \frac{C}{4}\frac{\phi^4 - \phi_e^4}{\phi_c^2}\right]; \quad \mathcal{N} \leq \mathcal{N}_f, \tag{6.40}$$

where $\mathcal{N}_f$ represents the value of $\mathcal{N}$ at the end of waterfall transition. For simplicity we assume that at the end of waterfall transition χ is very close to its local instantaneous minimum so χ^2 at $\mathcal{N} = \mathcal{N}_f$ is given by

$$\chi^2\left(n_f\right) = \chi_{min}^2\left(n_f\right) \simeq \frac{M^2}{\lambda} - \frac{g^2}{\lambda}\phi_f^2. \tag{6.41}$$

One can obtain $\mathcal{N}_f$ as a function of the number of e-folds from the critical epoch $\phi = \phi_f$ to the end of waterfall transition, n_f via

$$\mathcal{N}_f\left(n_f\right) = \frac{1}{4M_P^2}\left[\phi_f^2 - \phi_e^2 + \frac{C}{4\phi_c^2}\left(\phi_f^4 - \phi_e^4\right)\right], \tag{6.42}$$

from which one can easily find

$$4M_P^2 \delta \mathcal{N}_f (n_f) = \delta \left(\phi_f^2 \right) \left(1 + \frac{C}{2} \frac{\phi_f^2}{\phi_c^2} \right). \tag{6.43}$$

On the other hand, using Eq. (6.8), we obtain

$$-4M_P^2 \delta n_f \simeq \delta \left(\phi_f^2 \right) \left(1 + \frac{C}{2} \frac{\phi_c^2}{\phi_f^2} \right). \tag{6.44}$$

Using the relation $\phi_f^2 \simeq \phi_c^2 - 4M_P^2 n_f$ one finds that

$$\delta \mathcal{N}_f (n_f) \simeq -\delta n_f (1 - 2C\epsilon n_f) \quad \rightarrow \quad \frac{d\mathcal{N}_f}{dn_f} = -1 + 2C\epsilon n_f, \tag{6.45}$$

in which ϵ is the usual slow-roll parameter approximately given by $\epsilon \simeq 2M_P^2/\phi_c^2$.

Now we trace back the dynamics to earlier times before the end of waterfall transition when $\mathcal{N} > \mathcal{N}_f$. For this stage, instead of $\mathcal{N}$, it is more convenient to use n which is the number of e-folds counted *forward* in time from the critical point such that $n = n_f + \mathcal{N}_f - \mathcal{N}$. In this convention $\chi^2(n)$ is given by

$$\chi^2 (n) = \exp \left[2 \left(f(n) - f(n_f) \right) \right] \chi_{min}^2 (n_f), \tag{6.46}$$

where χ_{min} is given in Eq. (6.41) and for a sharp phase transition the function $f(n)$ is given by

$$f(n) = \frac{2}{3} \epsilon_\chi n^{3/2}. \tag{6.47}$$

During this stage, n is obtained in terms of $\phi(n)$ as given by Eq. (6.8),

$$-4M_P^2 n = \phi(n)^2 - \phi_c^2 \left[1 - C \ln \left(\frac{\phi}{\phi_c} \right) \right]. \tag{6.48}$$

It is worth noting that n depends on n_f and $\mathcal{N}$ non-trivially via

$$n(n_f, \mathcal{N}) = n_f + \mathcal{N}_f (n_f) - \mathcal{N}. \tag{6.49}$$

By the virtue of the above geometric relation and using Eq. (6.45) one obtains

$$\frac{\partial n}{\partial n_f} = 2C\epsilon n_f.$$

$$(6.50)$$

Keeping in mind the above dependence of n on n_f and $\mathcal{N}$, we can take the variation of Eqs. (6.46) and (6.48) yielding

$$\frac{\delta\chi^2(n)}{\langle\delta\chi^2(n)\rangle} = \frac{\delta\chi^2_{min}(n_f)}{\chi^2_{min}(n_f)} + 2f'(n)\delta n - 2f'(n_f)\delta n_f,$$

$$(6.51)$$

$$-2M_p^2\delta n \simeq \phi(n)\delta\phi(n)\left(1 + \frac{C}{2} + C\epsilon n\right).$$

$$(6.52)$$

On the other hand, from Eq. (6.41), we have

$$\chi^2_{min}(n_f) = 4M_P^2\frac{g^2}{\lambda}n_f,$$

$$(6.53)$$

where we made use of Eq. (6.8) and the fact that $\phi_f \lesssim \phi_c$. This yields

$$\frac{\delta\chi^2_{min}(n_f)}{\chi^2_{min}(n_f)} = \frac{\delta n_f}{n_f}.$$

$$(6.54)$$

Finally, solving Eqs. (6.51) and (6.52) for $\delta\mathcal{N}$, one obtains

$$\delta\mathcal{N} = \frac{\delta\chi^2(n)}{\langle\delta\chi^2(n)\rangle}\frac{\partial n}{\partial n_f}\frac{1}{-2f'(n_f) + n_f^{-1}}$$

$$+\frac{\phi\delta\phi(n)}{2M_P^2}\left[1 + \frac{C}{2} + C\epsilon n + \frac{2f'(n)}{-2f'(n_f) + n_f^{-1}}\frac{\partial n}{\partial n_f}\right].$$

$$(6.55)$$

To simplify the above expression, we note that $2n_f f'(n_f) \gg 1$ which is valid for a sharp phase transition. Using this approximation, the above expression simplifies to

$$\delta\mathcal{N} = -\frac{C\epsilon n_f}{f'(n_f)}\frac{\delta\chi^2(n)}{\langle\delta\chi^2(n)\rangle} + \left[1 + \frac{C}{2} + C\epsilon(n - 2n_f)\right]\frac{\phi\delta\phi}{2M_P^2}.$$

$$(6.56)$$

Noting that both $\delta\chi^2(n)$ and $\langle\chi^2(n)\rangle$ have the same n-dependence, the above formula can be expressed in terms of $\delta\chi^2(0)$ and $\langle\chi^2(0)\rangle$ as follows

$$\mathcal{R}_c = \delta\mathcal{N} = \left[1 + \frac{C}{2} + C\epsilon\left(n - 2n_f\right)\right]\frac{\phi\delta\phi}{2M_P^2} - \frac{C\epsilon n_f}{f'\left(n_f\right)}\frac{\delta\chi^2(0)}{\langle\delta\chi^2(0)\rangle},$$

(6.57)

where here and below we omit the subscript c from $\mathcal{R}_c$ for notational simplicity. We emphasize that the surface of end of inflation at $\phi = \phi_e$ is a "uniform energy density" hypersurface.

Equation (6.57) is our key formula for calculating the power spectrum. From the above formula we see that the curvature perturbation has the conventional contribution from the inflaton up to slow-roll corrections in the inflaton mass. In addition, we have the additional contribution from the waterfall field. The latter is a dynamical effect which is intrinsic to waterfall dynamics. This is in contrast to many other scenarios in which, as discussed in Sec. 6.1, the local features are added in *ad hoc* manner.

6.4   Power spectrum with localized feature

Having obtained δN and $\mathcal{R}_c$ to first order in fields perturbations, we now calculate the power spectrum. Our goal is to find the imprints of the sharp waterfall transition on the power spectrum.

As it is clear from Eq. (6.57), the total power spectrum can be divided into two distinct contributions, the contributions from the inflaton fluctuations and the contributions from the waterfall field fluctuations

$$\mathcal{P}_{\mathcal{R}_c} = \mathcal{P}_{\mathcal{R}_c}^{\phi} + \mathcal{P}_{\mathcal{R}_c}^{wf}$$
$$= \left[1 + \frac{C}{2} + \mathcal{O}(C\epsilon)\right]^2 \frac{\phi^4}{4M_P^4}\mathcal{P}_{\delta\phi/\phi} + \frac{C^2\epsilon^2 n_f^2}{f'^2\left(n_f\right)}\mathcal{P}_{\delta\chi^2/\chi^2}.$$

(6.58)

Below we calculate each contribution in turn.

6.4.1 Contribution of inflaton to power spectrum

Since the inflaton field is light throughout the whole period of inflation, the amplitude of its quantum fluctuations on initial flat hypersurface at the time of horizon crossing has the usual form

$$\delta\phi(k) = \frac{H(n_k)}{\sqrt{2k^3}}, \tag{6.59}$$

in which $H(n_k)$ represents the Hubble parameter at the time of horizon crossing. Using the above initial amplitude for $\delta\phi$ we obtain

$$\mathcal{P}^{\phi}_{\mathcal{R}_c}(k) \simeq \frac{1}{4\pi^2}\left[1 + \frac{C}{2}\right]^2 \frac{\phi^2 H^2}{4M_P^4}\bigg|_{n=n_k} \simeq \frac{(1+C)}{48\pi^2}\frac{\phi^2 V_{eff}^+(\phi)}{M_P^6}\bigg|_{n=n_k}, \tag{6.60}$$

where $V_{eff}^+(\phi)$ is given by Eq. (6.7), which to first order in C is given by

$$V_{eff}^+(\phi) = \frac{1}{2}m^2(1+C)\left(1 - \frac{C}{2}\frac{\phi^2}{\phi_c^2}\right)\phi^2 + \mathcal{O}(C\,\epsilon). \tag{6.61}$$

Plugging these results in power spectrum Eq. (6.60) yields

$$\mathcal{P}^{\phi}_{\mathcal{R}_c}(k) = \frac{1}{96\pi^2}(1+C)^2\left(1 - \frac{C}{2}\frac{\phi^2}{\phi_c^2}\right)\frac{m^2\phi^4}{M_P^6}\bigg|_{n=n_k}. \tag{6.62}$$

In the limit we neglect the corrections of $\mathcal{O}(C)$, the above expression reduces to the standard result, $\mathcal{P}^{\phi}_{\mathcal{R}_c}(k) = V^3/(12\pi^2 V'^2 M_P^6)|_{n=n_k}$.

6.4.2 Contribution of the waterfall field to power spectrum

Now we calculate the contribution of the waterfall field fluctuations to the power spectrum. This is somewhat non-trivial because, unlike the usual case (like the case of inflaton) in which the perturbations are linear in field perturbations, the contributions of $\delta\chi$ is quadratic, i.e. $\delta N \sim \delta\chi^2$. In order to calculate this contribution we first note that

$$\langle(\delta\chi^2)_{\mathbf{k}}(\delta\chi^2)_{\mathbf{q}}\rangle \equiv P_{\delta\chi^2}(k)(2\pi)^3\delta^3(\mathbf{k}+\mathbf{q}). \tag{6.63}$$

$$\mathcal{P}_{\delta\chi^2/\chi^2}(k) \equiv \frac{1}{\langle\delta\chi^2(0)\rangle^2}\frac{k^3}{2\pi^2}P_{\delta\chi^2}(k). \tag{6.64}$$

The correlation function of $\delta\chi_k^2$ can be calculated using the following identity [87],

$$\langle(\delta\chi^2)_k(\delta\chi^2)_q\rangle = 2\int \frac{d^3q}{(2\pi)^3}|\delta\chi_{|\mathbf{k}-\mathbf{q}|}|^2|\delta\chi_q|^2(2\pi)^3\delta^3(\mathbf{k}+\mathbf{q}) . \quad (6.65)$$

In Appendix C we demonstrate that

$$\mathcal{P}_{\delta\chi^2}(k) \simeq \frac{\xi^3 H^2}{4\pi^2}\mathcal{P}_{\delta\chi} , \quad (6.66)$$

in which ξ is a numerical factor of order unity. Consequently, this implies that the power spectrum of $\delta\chi^2$ is proportional to the power spectrum of $\delta\chi$. Plugging this result in Eqs. (6.58) and (6.64) and using the explicit forms of function $f(n)$ given in Eq. (6.47) and $\langle\delta\chi^2(0)\rangle$ obtained in Eq. (B.11) results in

$$\mathcal{P}_{\mathcal{R}_c}^{wf}(k) = \frac{4C^2\epsilon^2 n_f\xi^3}{3\epsilon_\chi^{10/3}}\frac{\mathcal{P}_{\delta\chi}(k;0)}{\langle\delta\chi^2(0)\rangle} . \quad (6.67)$$

Finally, plugging the spectrum of waterfall field perturbation (434) in the above expression yields

$$\mathcal{P}_{\mathcal{R}_c}^{wf}(k) \simeq \begin{cases} \frac{16}{9}C^2\epsilon^2 n_f\epsilon_\chi^{-20/3}\xi^3\left(\frac{k}{k_c}\right)^3 ; & k < k_c , \\ \frac{16}{9}C^2\epsilon^2 n_f\epsilon_\chi^{-8/3}\xi^3 n_t(k)\exp\left[-\frac{4}{3}\epsilon_\chi n_t^{3/2}(k)\right] ; & k > k_c . \end{cases}$$
$$\quad (6.68)$$

6.4.3   *Total curvature perturbation power spectrum*

Having calculated the contributions from the inflaton field and the waterfall field to power spectrum, we calculate the total power spectrum by adding them up. Since the waterfall contribution is localized, peaking at $k = k_{max} \gtrsim k_c$, let us compare the amplitudes of $\mathcal{P}_{\mathcal{R}_c}^{wf}(k)$ and $\mathcal{P}_{\mathcal{R}_c}^{\phi}(k)$ at $k \simeq k_{max}$. Using Eq. (6.33) for k_{max}, from Eqs. (6.68) and (6.62) we obtain

$$\frac{\mathcal{P}_{\mathcal{R}_c}^{wf}(k_{max})}{\mathcal{P}_{\mathcal{R}_c}^{\phi}} \simeq 10^3 C^2\left(\frac{\epsilon}{10^{-2}}\right)^4\left(\frac{\epsilon_\chi}{10}\right)^{-10/3} , \quad (6.69)$$

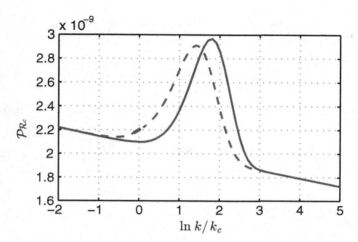

Fig 6.2 Power spectrum of the curvature perturbation. The red dashed curve represents an analytical estimate of the total curvature perturbation power from the contributions of Eqs. (6.62) and (6.68). The blue solid curve represents the total curvature perturbation, $\mathcal{P}_{\mathcal{R}_c}(k)$, by calculating the the convolution integral Eq. (6.65) numerically. As we see, $\mathcal{P}_{\mathcal{R}_c}^{wf}(k)$ peaks near $k_{max} \sim k_c$ and decays quickly for k beyond k_{max}. The parameters are the same as in Fig. 6.1. The figure is borrowed from Ref. [84].

where we have used $m/M_P \sim 10^{-6}$ in order to satisfy the COBE normalization. This ratio indicates that there can be a prominent peak in power spectrum even for a small value of C, say $C \sim 0.1$.

In Fig. 6.2 we have plotted the total curvature perturbation power spectrum for the parameters $C = 0.15$, $\epsilon = 0.01$ and $\epsilon_\chi = 20$. The localized peak at $k = k_{max} \sim k_c$ is from the waterfall field contribution, $\mathcal{P}_{\mathcal{R}_c}^{wf}(k)$. For the region away from the location of the peak the spectrum is dominated by the inflaton contribution, $\mathcal{P}_{\mathcal{R}_c}^{\phi}(k)$, similar to standard chaotic inflation.

It is instructive to estimate n_f, the duration of the waterfall phase transition. As explained before, we take $\langle \delta\chi^2 \rangle$ as the classical value of the waterfall field averaged on each horizon size patch. A good condition for the completion of the phase transition is when

$\langle \delta\chi^2 \rangle$ reaches its local minimum $\chi^2_{min}(n_f)$ given by Eq. (6.41). Using Eq. (6.48), we obtain

$$\langle \delta\chi^2 (n_f) \rangle \simeq \frac{4g^2 M_P^2}{\lambda} n_f. \tag{6.70}$$

In Appendix B the expectation value $\langle \delta\chi^2(n) \rangle$ is obtained to be

$$\langle \delta\chi^2 (n) \rangle = \langle \delta\chi^2 (0) \rangle \exp\left(\frac{4}{3}\epsilon_\chi n^{3/2}\right) \simeq \frac{3\epsilon_\chi^{4/3} H^2}{16\pi^2} \exp\left(\frac{4}{3}\epsilon_\chi n^{3/2}\right). \tag{6.71}$$

Equating this with $\chi^2_{min}(n_f)$, the following estimate for n_f is obtained

$$n_f = \Gamma \epsilon_\chi^{-2/3}; \quad \Gamma \simeq \left(\ln\left[\frac{32\pi^2 \epsilon_\chi^{2/3}}{6\lambda}\right]\right)^{2/3}. \tag{6.72}$$

For the numerical example specified above, we obtain $n_f \simeq 0.5$ so the phase transition is fairly sharp. However, we note that it is smooth enough to keep the dynamics of the phase transition adiabatic. In other words, the adiabaticity condition of the inflaton vacuum state is not violated, whereas for an infinitely sharp transition the adiabaticity condition may no longer hold.

The bispectrum analysis of this setup is more complicated, so we have relegated the corresponding δN analysis to Appendix D where the interested reader is invited for detail studies of higher order δN analysis.

To summarize, in this chapter we have presented yet another application of δN formalism in a model in which the perturbations are non-Gaussian in nature, i.e. $\mathcal{R}_c \sim \delta\chi^2$ induced from the waterfall transition. This in turn generates spikes in power spectrum and bispectrum. The model is similar to the hybrid scenario, but in contrast to hybrid model, the vacuum energy of the waterfall field is sub-dominant compared to inflaton mass term. We choose the model parameters such that the waterfall transition is sharp, i.e. the waterfall transition takes place in about one e-fold or so. Since the waterfall is coupled to inflaton via interaction $g^2\phi^2\chi^2$, the waterfall transition also induces a sharp yet small change in the inflaton mass. As we have seen the waterfall quantum fluctuations play crucial roles in determining the local feature. Because of the non-Gaussian nature of the $\delta\chi^2$ fluctuations, the power spectrum has the convoluted form $\mathcal{P}_{\mathcal{R}_c} \sim \langle (\delta\chi^2)(\delta\chi^2) \rangle$ while the bispectrum has the form of $B_{\mathcal{R}_c} \sim \langle (\delta\chi^2)(\delta\chi^2)(\delta\chi^2) \rangle$.

Since the waterfall field is very massive, it has no classical evolution during the first stage of inflation (before the waterfall transition) and it is locked in its local minimum $\chi = 0$. However, once ϕ reaches a critical value, χ becomes tachyonic and the waterfall transition is triggered. Consequently, the squared fluctuation, $\delta\chi^2$, grows exponentially. We have argued that $\langle \delta\chi^2 \rangle$ averaged over each Hubble horizon patch determines the effective classical trajectory. The fluctuations from one horizon patch to another is given by $\Delta\chi^2 = \delta\chi^2 - \langle \delta\chi^2 \rangle$, which takes into account the fluctuation around the classical trajectory constructed by $\langle \delta\chi^2 \rangle$.

We have calculated the power spectrum of $\Delta\chi^2$ fluctuations. Because of the assumption of a sharp waterfall transition, the induced power spectrum from $\Delta\chi^2$ fluctuations peaks near the comoving scale k_c which crosses the horizon at the time of waterfall transition, $k = k_{max} \sim k_c$. Depending on the model parameters, the induced power spectrum can be comparable or larger than the contribution from the inflaton field. As we argued in Sec. 6.1, this local feature may be used to explain glitches found in the observed CMB angular power spectrum. In this simple toy model we have considered a single waterfall transition. However, one may allow many waterfall fields coupled to the inflaton to generate a series of waterfall phase transitions which can induce multiple local features in power spectrum and bispectrum.

Because of the intrinsic non-Gaussian nature of $\delta\chi^2$ distribution, one expects to obtain large spiky non-Gaussianities from waterfall field dynamics. As we have shown in Appendix D, f_{NL} has both intrinsic and dynamical contributions. The intrinsic contributions in f_{NL} comes from the $\delta\chi^2$ three-point function while the dynamical part originates from the non-linear dynamics of the waterfall field. We have shown that $f_{NL}(k_{max}) \sim \epsilon_\chi/\epsilon C$, in which ϵ_χ is a measure of the sharpness of the waterfall transition. Consequently, the sharper is the transition (i.e. the larger is ϵ_χ) the larger is f_{NL}. Similar to the case of power spectrum, the bispectrum peaks narrowly at $k \simeq k_{max}$ with possible large value of f_{NL}. Therefore, one has to perform an independent CMB data analysis to see if this spiky non-Gaussianity is consistent with observations.

APPENDIX A

δN for general c_S in non-attractor background

Having presented the δN analysis for the simple case $c_s = 1$, here we present the δN analysis for arbitrary value of c_s in non-attractor background in Chapter 5.

We are in the limit in which the term linear in X can be neglected in $P(X, \phi)$. The scalar field equation of motion is

$$\ddot{\phi} + 3c_s^2 H \dot{\phi} - F = 0, \qquad (A.1)$$

where we have defined

$$F \equiv c_s^2 \frac{P_{,\phi}}{P_{,X}}. \qquad (A.2)$$

It is not easy to find the general solution to this equation. Let us thus first consider our particular solution given by Eq. (5.8)

$$\phi = \phi_0 = \phi_* e^{\kappa H t} = \phi_* e^{-\kappa N}, \qquad (A.3)$$

and then obtain a more general solution for the background up to the second order in perturbations around this particular solution. As before, it is assumed that the non-attractor phase ends when $\phi = \phi_*$.

Plugging the special solution ϕ_0 given in Eq. (A.3) in F yields

$$F = c_s^2 \frac{P_{,\phi}}{P_{,X}} = F_0 \left(\frac{\phi}{\phi_0}\right)^{2\alpha-1} \left(\frac{\dot{\phi}}{\dot{\phi}_0}\right)^{2-2\alpha}, \qquad (A.4)$$

in which

$$F_0 \equiv \ddot{\phi}_0 + 3c_s^2 H \dot{\phi}_0 = \kappa \left(\kappa + 3c_s^2\right) H^2 \phi_0. \qquad (A.5)$$

Let us expand F around the special solution $\phi = \phi_0(N)$ to the second order. For notational simplicity let us define

$$\chi \equiv \phi - \phi_0.$$

Plugging this in F and expanding to second order yields

$$\begin{aligned}
F = F_0 &\left[1 + (2 - 2\alpha)\frac{\dot{\chi}}{\dot{\phi}_0} + (2\alpha - 1)\frac{\chi}{\phi_0} + \frac{(2 - 2\alpha)(1 - 2\alpha)}{2}\left(\frac{\dot{\chi}}{\dot{\phi}_0}\right)^2 \right. \\
&\left. + (2 - 2\alpha)(2\alpha - 1)\frac{\dot{\chi}\chi}{\dot{\phi}_0 \phi_0} + \frac{(2\alpha - 1)(2\alpha - 2)}{2}\left(\frac{\chi}{\phi_0}\right)^2\right] \\
= F_0 &\left[1 + \frac{(2 - 2\alpha)}{\kappa}\frac{\dot{\chi}}{H\phi_0} + (2\alpha - 1)\frac{\chi}{\phi_0} + \frac{(2 - 2\alpha)(1 - 2\alpha)}{2\kappa^2}\left(\frac{\dot{\chi}}{H\phi_0}\right)^2 \right. \\
&\left. + \frac{(2 - 2\alpha)(2\alpha - 1)}{\kappa}\frac{\dot{\chi}\chi}{H\phi_0^2} + \frac{(2\alpha - 1)(2\alpha - 2)}{2}\left(\frac{\chi}{\phi_0}\right)^2\right].
\end{aligned}$$
$$(A.6)$$

With F obtained to second order in χ perturbations, we can solve Eq. (A.1) perturbatively as follows.

Linear perturbation

To start, let us consider the linear perturbation, $\chi = \chi_1$. The equation of motion is

$$0 = \ddot{\chi}_1 + 3c_s^2 H \dot{\chi}_1 - F_0 \left[\frac{(2 - 2\alpha)}{\kappa} \frac{\dot{\chi}_1}{H\phi_0} + (2\alpha - 1)\frac{\chi_1}{\phi_0} \right]$$

$$= \ddot{\chi}_1 + [3c_s^2 + (2\alpha - 2)(\kappa + 3c_s^2)] H \dot{\chi}_1 - (2\alpha - 1)\kappa(\kappa + 3c_s^2)H^2\chi_1$$

$$= \ddot{\chi}_1 + [3 + (2\alpha - 1)\kappa - \kappa] H \dot{\chi}_1 - \kappa [3 + (2\alpha - 1)\kappa] H^2\chi_1.$$

$$(A.7)$$

The general solution is obtained to be

$$\chi = \chi_1 \propto \begin{cases} \exp\ [\kappa H t], \\ \exp\ [-(3 + \eta - \kappa)Ht], \end{cases} \quad (A.8)$$

in which the relations $2\alpha\kappa = \eta$ and $\eta = \dot{\epsilon}/H\epsilon$ have been used. As we have seen, a scale-invariant spectrum requires $\eta \simeq -6$, so the second solution scaling as $\exp\ [-(3 + \eta - \kappa)Ht]$ is the dominant solution.

Second-order perturbation

Next we look at the second-order perturbation, χ_2. The equation of motion is

$$\ddot{\chi}_2 + [3 + (2\alpha - 1)\kappa - \kappa] H \dot{\chi}_2 - \kappa[3 + (2\alpha - 1)\kappa]H^2\chi_2 = S, \quad (A.9)$$

in which the source term S is defined by

$$S \equiv F_0 \left[\frac{(2 - 2\alpha)(1 - 2\alpha)}{2\kappa^2} \left(\frac{\dot{\chi}_1}{H\phi_0} \right)^2 + \frac{(2 - 2\alpha)(2\alpha - 1)}{\kappa} \frac{\dot{\chi}_1 \chi_1}{H\phi_0^2} \right.$$

$$\left. + \frac{(2\alpha - 1)(2\alpha - 2)}{2} \left(\frac{\chi_1}{\phi_0} \right)^2 \right].$$

$$(A.10)$$

As explained above, the dominant solution for χ_1 is $\chi_1 \propto \exp\ [-(3 + \eta - \kappa)Ht]$. With this solution for χ_1 we find that $\chi_2 \propto e^{\mu H t}$ is a solution to Eq. (A.9) in which μ is determined from the time dependence of S. We thus obtain

$$\chi_2 = \left(\frac{g}{\phi_0}\right)\chi_1^2, \tag{A.11}$$

with

$$g = \frac{(3c_s^2 + \kappa)}{4\kappa}(2 - 2\alpha)(1 - 2\alpha), \tag{A.12}$$

and

$$\mu = -2(3 + \eta) + \kappa. \tag{A.13}$$

Calculating δN

We are ready to calculate δN to second order in field perturbations. The general background solution of ϕ, computed up to the second order in field perturbations around the reference trajectory $\phi_0 \propto e^{-\kappa N}$, are given by Eqs. (A.8) and (A.11). Combined together, we have

$$\phi = \phi_0(N) + \chi_1 + \left(\frac{g}{\phi_0}\right)\chi_1^2, \tag{A.14}$$

in which $\chi_1 \propto \exp\left[-(3 + \eta - \kappa)Ht\right] = e^{(3+\eta-\kappa)N}$. Expressed in terms of $N = -Ht$, this yields

$$\begin{aligned}
\phi &= \phi_{0*}\left(e^{-\kappa N} + \lambda e^{(3+\eta-\kappa)N} + g\lambda^2 e^{(2(3+\eta)-\kappa)N}\right) \\
&= \frac{\phi_*}{1 + \lambda + g\lambda^2}\left(e^{-\kappa N} + \lambda e^{(3+\eta-\kappa)N} + g\lambda^2 e^{(2(3+\eta)-\kappa)N}\right),
\end{aligned} \tag{A.15}$$

where λ is an integration constant that parametrizes different trajectories and we have set $\phi(0, \lambda) = \phi_*$ for any value of λ. This comes from the assumption that the end of the non-attractor phase is determined only by the value of the scalar field, $\phi = \phi_*$. More specifically, $\phi_* = \phi_{0*} + \chi_{1*} + \chi_{2*} = \phi_{0*}(1 + \lambda + g\lambda^2)$ and

$$\lambda = \chi_{1*}/\phi_{0*}. \tag{A.16}$$

This indicates that λ itself is first order in perturbations. Note that we use the obvious definition that $\chi_{1*} = \chi_1(N = 0)$, i.e. χ_{1*} is the value of χ_1 at the end of non-attractor phase with similar interpretation for ϕ_{0*} and χ_{2*}.

In principle Eq. (A.15) can be inverted to give N as a function of ϕ and λ. However, that is not necessary for our purpose. All we need is to perturb both sides of Eq. (A.15) to second order in perturbation as $\phi_* = \phi_{0*} + \delta\phi_{1*} + \delta\phi_{2*}$ and $N = 0 + \delta N_1 + \delta N_2$ and equate both sides accordingly. Note that our convention is such that at the end of non-attractor phase $N = 0$ so that is why the leading term in N starts with zero. With this prescription in mind and noting that λ, as given in Eq. (A.16), is first order in perturbations, we get

$$\delta N_1 = -\frac{\lambda}{\kappa} = -\frac{\chi_{1*}}{\kappa\phi_*}, \tag{A.17}$$

and

$$\delta N_2 = -\frac{1}{\kappa\phi_*}\left[\delta\phi_{2*} - \frac{1}{2\phi_*}\delta\phi_{1*}^2 + \frac{3+\eta}{\kappa}\delta\phi_{1*}\lambda\right], \tag{A.18}$$

with $\delta\phi_1$ and $\delta\phi_2$ being the first and second order perturbations around the reference trajectory, ϕ_0. Recalling $\delta\phi_1 = \chi_1$ and $\delta\phi_2 = \chi_2 = g\chi_1^2/\phi_0$, we rewrite Eq. (A.18) as

$$\delta N_2 = -\frac{1}{\kappa\phi_*^2}\left(g - \frac{1}{2} + \frac{\eta+3}{\kappa}\right)\chi_{1*}^2. \tag{A.19}$$

We thus finally obtain the total δN calculated at $N = 0$ as

$$\delta N = \delta N_1 + \delta N_2$$
$$= -\frac{\chi_{1*}}{\kappa\phi_*} - \left(g - \frac{1}{2} + \frac{\eta+3}{\kappa}\right)\frac{\chi_{1*}^2}{\kappa\phi_*^2}. \tag{A.20}$$

This is the main result of this section.

Having obtained δN to second order in perturbations, we can calculate the amplitude of local-type non-Gaussianity. Comparing with the definition of f_{NL}^{local} given in Eq. (5.41) we obtain

$$
\begin{aligned}
\frac{3}{5} f_{NL}^{\text{local}} &= -\kappa g + \frac{\kappa}{2} - (3 + \eta) \\
&= \frac{(3 + \eta + 3c_s^2)}{4\,(1 + c_s^2)} \left(1 - \frac{1}{c_s^2}\right) - (3 + \eta) + \frac{\eta c_s^2}{2\,(1 + c_s^2)} \qquad \text{(A.21)} \\
&= -\frac{1}{4c_s^2} \left(9c_s^2 + 2\eta c_s^2 + 3 + \eta\right).
\end{aligned}
$$

With $\eta = -6$ in order to obtain a scale-invariant power spectrum, f_{NL}^{local} simplifies to

$$
f_{NL}^{\text{local}} = \frac{5}{4c_s^2} \left(1 + c_s^2\right). \qquad \text{(A.22)}
$$

This result is valid for any values of c_s because we have not assumed $c_s \ll 1$ during our analysis. We see that when $c_s = 1$ we obtain $f_{NL}^{\text{local}} = 5/2$ which agrees with Eq. (5.42).

APPENDIX B

Variance of $\delta\chi$ fluctuations

In this appendix we calculate $\langle\delta\chi^2(n)\rangle$ in Chapter 6. As we argued in the main text, $\langle\delta\chi^2(n)\rangle$, averaged over a Hubble patch, represents the background trajectory after it begins to evolve as a classical field and

$$\sqrt{\langle\delta\chi^2(n)\rangle} \propto \exp\left[\frac{2}{3}\epsilon_\chi n^{3/2}\right]. \tag{B.1}$$

It is convenient to define $\langle\delta\chi^2(0)\rangle$ not by its actual value at $n = 0$, but by the value it would have taken if it evolved classically from the beginning. In other words, for modes which become tachyonic by the end of the waterfall transition $n = n_f$, we define

$$\langle\delta\chi^2(0)\rangle \equiv \langle\delta\chi^2(n_f)\rangle \exp\left[-\frac{4}{3}\epsilon_\chi n_f^{3/2}\right]. \tag{B.2}$$

The small scales and the large scales contribute differently in $\langle\delta\chi^2(0)\rangle$ so we separate their contributions denoted by subscript S and L, respectively,

$$\langle\delta\chi^2(0)\rangle = \langle\delta\chi^2(0)\rangle_S + \langle\delta\chi^2(0)\rangle_L. \tag{B.3}$$

It is shown in Ref. [87] that for the large scale we have

$$\langle\delta\chi^2(0)\rangle_L = \frac{H^2}{4\pi^2\epsilon_\chi}. \tag{B.4}$$

145

The contributions of the small scales were also studied in the past in which, as we show below, it was somewhat overestimated. Here we present a more accurate estimation as follows.

As discussed in the main text around Eq. (6.28), one approximately has

$$\delta\chi_k^S(n) = \frac{H}{\sqrt{2k}\,k_c}e^{-n_t}\exp\left[\frac{2}{3}\epsilon_\chi\left(n^{3/2} - n_t^{3/2}\right)\right]; \quad n > n_t(k). \quad (\text{B.5})$$

However, the above result does not reproduce the correct behavior when $n - n_t \ll 1$. In this limit, it behaves as (in a slightly different notation)

$$\delta\overline{\chi}_k^S(n) = \frac{H}{\sqrt{2k}\,k_c}e^{-n_t}\exp\left[\frac{2}{3}\epsilon_\chi(n - n_t)^{3/2}\right]. \quad (\text{B.6})$$

If we extrapolate this result to $n = n_f$, we obtain

$$\delta\overline{\chi}_k^S(n_f) = \frac{H}{\sqrt{2k}\,k_c}e^{-n_t}\exp\left[\frac{2}{3}\epsilon_\chi(n_f - n_t)^{3/2}\right], \quad (\text{B.7})$$

which is different from the one that follows from the approximate formula (B.5),

$$\delta\chi_k^S(n_f) = \frac{H}{\sqrt{2k}\,k_c}e^{-n_t}\exp\left[\frac{2}{3}\epsilon_\chi\left(n_f^{3/2} - n_t^{3/2}\right)\right]. \quad (\text{B.8})$$

Consequently, Eq. (B.7) provides an estimate for $\delta\chi_k^S(0)$ as follows

$$\begin{aligned}\delta\overline{\chi}_k^S(0) &= \frac{H}{\sqrt{2k}\,k_c}e^{-n_t}\exp\left[\frac{2}{3}\epsilon_\chi(n_f - n_t)^{3/2} - \epsilon_\chi n_f^{3/2}\right]\\ &= \delta\chi_k^S(0)\exp\left[\frac{2}{3}\epsilon_\chi(n_f - n_t)^{3/2} - \epsilon_\chi\left(n_f^{3/2} - n_t^{3/2}\right)\right],\end{aligned} \quad (\text{B.9})$$

in which Eq. (B.5) have been used to estimate $\delta\chi_k^S(0)$. One can easily check that $\delta\overline{\chi}_k^S(0) \le \delta\chi_k^S(0)$. Thus one concludes that using $\delta\overline{\chi}_k^S(0)$ yields a slight underestimate, while using $\delta\chi_k^S(0)$ results in a slight overestimate.

Let us estimate $\langle\delta\chi^2(0)\rangle_S$ obtained from $\delta\chi_k^S(0)$. Using Eq. (6.25), one obtains

$$\langle\delta\chi^2(0)\rangle_S = \frac{\epsilon_\chi^2 H^2}{4\pi^2} \int_{n_t=0}^{n_f} dn_t \left(n_t + \frac{1}{2}\right) e^{-4/3\epsilon_\chi n_t^{3/2}}. \tag{B.10}$$

The exponential form of the integrand function suggests a natural cut-off at $n_{cut} \sim \epsilon_\chi^{-2/3}$. Therefore, we can neglect n_t compared to $1/2$ in the integrand which enables us to extend the range of integral to infinity, yielding

$$\begin{aligned}
\langle\delta\chi^2(0)\rangle_S &\simeq \frac{\epsilon_\chi^2 H^2}{8\pi^2} \int_{n_t=0}^{n_f} dn_t\, e^{-4/3\epsilon_\chi n_t^{3/2}} \\
&= \frac{\epsilon_\chi^2 H^2}{8\pi^2} \frac{\Gamma(\frac{2}{3})}{6^{1/3}\epsilon_\chi^{2/3}} \\
&\simeq 0.75\frac{\epsilon_\chi^{4/3} H^2}{8\pi^2}.
\end{aligned} \tag{B.11}$$

As mentioned above, this gives a slight overestimate of the true value of $\langle\delta\chi^2(0)\rangle_S$.

Now let us calculate the variance of $\delta\overline{\chi}_k^S(0)$. The integral in Eq. (B.10) is replaced with

$$\begin{aligned}
\langle\delta\overline{\chi}^2(0)\rangle_S &= \frac{\epsilon_\chi^2 H^2}{4\pi^2} \int_{n_t=0}^{n_f} dn_t \left(n_t + \frac{1}{2}\right) \\
&\times \exp\left[\frac{4}{3}\epsilon_\chi n_f^{3/2}\left((1 - \tfrac{n_t}{n_f})^{3/2} - 1\right)\right].
\end{aligned} \tag{B.12}$$

Now, as in the previous case, the integral is dominated by the integrand at $n_t/n_f \ll 1$, so by expanding the exponent in n_t/n_f, one can approximate it to obtain

$$\begin{aligned}
\langle\delta\overline{\chi}^2(0)\rangle_S &\simeq \frac{\epsilon_\chi^2 H^2}{8\pi^2} \int_{n_t=0}^{n_f} dn_t \exp\left[-2\epsilon_\chi n_f^{1/2} n_t\right] \\
&\simeq \frac{\epsilon_\chi^2 H^2}{8\pi^2} \frac{1}{2\epsilon_\chi n_f^{1/2}} \\
&= \frac{\epsilon_\chi H^2}{16\pi^2 n_f^{1/2}}.
\end{aligned} \tag{B.13}$$

This result is smaller than the result in (B.11) by a factor $\epsilon_\chi^{1/3} n_f^{-1/2}$. However, for reasonable values of ϵ_χ and n_f, say $\epsilon_\chi \sim 10$ and $n_f \sim 0.5$, this factor is around $\epsilon_\chi^{1/3} n_f^{-1/2} \sim 1.5$. Therefore, there is not a big difference between the above two estimates.

To summarize, we have shown that the dominant contribution to $\langle \delta\chi^2(0) \rangle_S$ comes from the modes around $\epsilon_\chi^{-1} n_f^{-1/2} \lesssim n \lesssim \epsilon_\chi^{-2/3}$, and is given approximately by either Eq. (B.11) or (B.13). Comparing these with Eq. (B.4), we find that the dominant contribution to the total variance of the waterfall field fluctuations comes from the small scale perturbations and

$$\langle \delta\chi^2(0) \rangle = \langle \delta\chi^2(0) \rangle_S + \langle \delta\chi^2(0) \rangle_L \simeq \langle \delta\chi^2(0) \rangle_S, \tag{B.14}$$

in which $\langle \delta\chi^2(0) \rangle_S$ is given by either Eq. (B.11) or (B.13), which are nearly equal to each other in our approximation in which the waterfall transition is sharp.

APPENDIX C

Correlation functions of $\delta\chi^2$

In this appendix we try to find a good approximation for the correlation functions of the $\delta\chi^2$ appearing in the power spectrum and bispectrum analysis, Eqs. (6.66) and (D.21).

First, we start with the following convolution integral which is necessary in order to calculate the power spectrum of the waterfall field

$$\langle(\delta\chi^2)_k (\delta\chi^2)_q\rangle = 2\int \frac{d^3q}{(2\pi)^3} |\delta\chi_{|\mathbf{k}-\mathbf{q}|}|^2 |\delta\chi_q|^2 \, (2\pi)^3\delta^3(\mathbf{k}+\mathbf{q}). \quad \text{(C.1)}$$

Decomposing the above integral into the radial and angular parts we obtain

$$2\int \frac{d^3q}{(2\pi)^3} |\delta\chi_{|\mathbf{k}-\mathbf{q}|}|^2 |\delta\chi_q|^2 = \int dn_q \mathcal{P}_{\delta\chi}(q) \int d(-\cos\theta) |\delta\chi_{|\mathbf{k}-\mathbf{q}|}|^2. \quad \text{(C.2)}$$

The above integral cannot be calculated analytically but we can find an estimate for it as follows. First, note that the power spectrum of waterfall field $\mathcal{P}_{\delta\chi}$ is highly peaked around $k = k_{max}$, so one expects that the curvature perturbation $\mathcal{P}_{\mathcal{R}} \sim \mathcal{P}_{\delta\chi^2}$ is also peaked near $k_{max} \gg k_c$. We will check this assumption in the following. Another important point is that for $|\mathbf{k}-\mathbf{q}| < k_c$, Eq. (6.32) shows that

$|\delta\chi_{|\mathbf{k}-\mathbf{q}|}|^2$ is nearly constant equal to $\frac{H_0^2}{2k_c^3}$, while it falls off exponentially for $|\mathbf{k}-\mathbf{q}| > k_c$. Therefore, we conclude that the second integral in Eq. (C.2) is negligible except for the interval

$$|\mathbf{k}-\mathbf{q}| \lesssim \xi k_c, \qquad (C.3)$$

in which the numerical factor ξ can be calculated using Eq. (6.29). To find an estimate for ξ it is natural to estimate the width Δk, the point at which $P_{\delta\chi}$ is reduced by a factor $1/e$ compared to its value at $k = k_c$, $P_{\delta\chi}(k_c) = H_0^2/(2k^3)$. Now using Eq. (6.29) and noting that for $k \sim k_c$, $n_t(k) \ll n_*(k)$, one concludes that

$$\xi \simeq e. \qquad (C.4)$$

In order to satisfy condition (529) one finds that the amplitude of q should be nearly equal to k and at the same time the angle between $\mathbf{k}$ and $\mathbf{q}$, denoted by θ, should be nearly zero. Now defining

$$q = k + \Delta q, \qquad \Delta q < k, \qquad \text{and} \qquad \theta = 0 + \Delta\theta \qquad (C.5)$$

one concludes that the condition

$$|\mathbf{k}-\mathbf{q}|^2 \lesssim \xi^2 k_c^2 \qquad (C.6)$$

yields

$$\Delta q \lesssim \xi k_c \quad \rightarrow \quad \Delta n_q \lesssim \frac{\xi k_c}{k}, \qquad (C.7)$$

$$\Delta(-\cos\theta) = \frac{\xi^2 k_c^2}{2k^2}. \qquad (C.8)$$

With these approximations we can estimate the integral in Eq. (C.2) for $k \gg k_c$ obtaining

$$2\int \frac{d^3q}{(2\pi)^3}|\delta\chi_{|\mathbf{k}-\mathbf{q}|}|^2|\delta\chi_q|^2 = 2\int dn_q d(-\cos\theta)|\delta\chi_{|\mathbf{k}-\mathbf{q}|}|^2 \mathcal{P}_{|\delta_\chi|}(q)$$

$$\simeq \Delta n_q \Delta(-\cos\theta)\frac{H_0^2}{2k_c^3}\mathcal{P}_{\delta_\chi}(k)$$

$$= \frac{\xi^3 H_0^2}{2k^3}\mathcal{P}_{\delta_\chi}(k). \qquad (C.9)$$

The above result yields the following good approximate for $\mathcal{P}_{\delta\chi^2}$ in terms of $\mathcal{P}_{\delta\chi}$:

$$\mathcal{P}_{\delta\chi^2}(k) \simeq \frac{\xi^3 H_0^2}{4\pi^2} \mathcal{P}_{\delta\chi}(k). \tag{C.10}$$

Similarly, we calculate approximately the three-point correlation function of $\delta\chi^2$ in the squeezed limit

$$\langle (\delta\chi^2)_{\mathbf{k}_1} (\delta\chi^2)_{\mathbf{k}_2} (\delta\chi^2)_{\mathbf{k}_3} \rangle_{\mathrm{sq}}$$
$$\simeq 8\,\delta^3 (\mathbf{k_1} + \mathbf{k_2} + \mathbf{k_3}) \int d^3q |\delta\chi_q|^2 |\delta\chi_{|\mathbf{k}-\mathbf{q}|}|^4. \tag{C.11}$$

Following the same strategy as in the case of two-point correlation function, we can approximate the above integral obtaining

$$\langle (\delta\chi^2)_{\mathbf{k}_1} (\delta\chi^2)_{\mathbf{k}_2} (\delta\chi^2)_{\mathbf{k}_3} \rangle_{\mathrm{sq}} \simeq 8(2\pi)^2 \delta^3 (\mathbf{k_1} + \mathbf{k_2} + \mathbf{k_3}) \int dn_q \mathcal{P}_{\delta\chi}(q)$$
$$\int d(-\cos\theta) |\delta\chi_{|\mathbf{k}-\mathbf{q}|}|^4. \tag{C.12}$$

Now using the above relation we obtain

$$B_{\delta\chi^2}^{\mathrm{sq}} = 8 \int dn_q \mathcal{P}_{\delta\chi}(q) \int d(-\cos\theta) |\delta\chi_{|\mathbf{k}-\mathbf{q}|}|^4$$
$$\simeq 8\Delta n_q \Delta(-\cos\theta) \frac{H_0^4}{4k_c^6} \mathcal{P}_{\delta\chi}(k) \tag{C.13}$$
$$= \frac{\xi'^3 H_0^4}{k_c^3 k^3} \mathcal{P}_{\delta\chi}(k),$$

in which the numerical factor ξ', similarly defined as ξ, is approximately given by $\xi' \simeq \sqrt{e}$. Alternatively, one can also write the above result as

$$B_{\delta\chi^2}^{\mathrm{sq}}(k) \simeq \frac{\xi'^3}{k_c^3} \frac{H_0^4}{2\pi^2} \mathcal{P}_{\delta\chi}(k). \tag{C.14}$$

APPENDIX **D**

Bispectrum with localized feature

In this appendix we present the bispectrum analysis of the curvature perturbations with localized feature in Chapter 6. Because of the intrinsic non-Gaussian nature of $\delta\chi^2$, it is natural to expect large spiky non-Gaussianities when $\mathcal{P}_{\mathcal{R}_c}^{wf}(k_{max}) > \mathcal{P}_{\mathcal{R}_c}^{\phi}(k_{max})$.

In the previous analysis, leading to Eq. (6.58), we have calculated δN up to $\delta\chi^2$. In order to calculate the bispectrum we have to expand δN up to $\delta\chi^4$. This analysis is presented in Appendix E. Using Eq. (E.3) the three-point function is obtained to be

$$
\langle \mathcal{R}_c(\mathbf{k}_1)\, \mathcal{R}_c(\mathbf{k}_2)\, \mathcal{R}_c(\mathbf{k}_3) \rangle
$$

$$
\equiv B_{\mathcal{R}_c}(\mathbf{k}_1, \mathbf{k}_2, \mathbf{k}_3)\, (2\pi)^3 \delta(\mathbf{k}_1 + \mathbf{k}_2 + \mathbf{k}_3)
$$

$$
= (N_{,\chi^2})^3 \left\langle (\delta\chi^2)_{\mathbf{k}_1} (\delta\chi^2)_{\mathbf{k}_2} (\delta\chi^2)_{\mathbf{k}_3} \right\rangle
$$

$$
+ \frac{1}{2}(N_{,\chi^2})^2 N_{,\chi^2,\chi^2} \left\langle [(\Delta\chi^2)^2]_{\mathbf{k}_1} (\delta\chi^2)_{\mathbf{k}_2} (\delta\chi^2)_{\mathbf{k}_3} + \text{c.p.} \right\rangle
$$

$$
+ \frac{1}{2}(N_{,\phi})^2 N_{,\phi\phi} \langle (\delta\phi^2)_{\mathbf{k}_1} \delta\phi_{\mathbf{k}_2} \delta\phi_{\mathbf{k}_3} + \text{c.p.} \rangle ,
$$

$$
\tag{D.1}
$$

where c.p. indicates the cyclic permutations i.e. $(\mathbf{k}_1, \mathbf{k}_2, \mathbf{k}_3) \rightarrow (\mathbf{k}_2, \mathbf{k}_3, \mathbf{k}_1) \rightarrow (\mathbf{k}_3, \mathbf{k}_1, \mathbf{k}_2)$ and $\Delta\chi^2$ represents the fluctuations of $\delta\chi^2(n, \mathbf{x})$ on superhorizon scales as defined in Eq. (6.13).

There are two different types of contributions to the three-point correlation function. The first term in Eq. (D.1) originates from the intrinsic non-Gaussianity of $\delta\chi^2$. We define the corresponding intrinsic bispectrum of $\delta\chi^2$ in the usual way via

$$\left\langle (\delta\chi^2)_{\mathbf{k}_1}(\delta\chi^2)_{\mathbf{k}_2}(\delta\chi^2)_{\mathbf{k}_3} \right\rangle = B_{\delta\chi^2}(k_1,k_2,k_3)(2\pi)^3\delta^3(\mathbf{k}_1+\mathbf{k}_2+\mathbf{k}_3).$$
(D.2)

The second term in Eq. (D.1) comes from the non-linear dynamics of the waterfall field. Finally, the last term represents the contribution of the inflaton field which generally is negligible when inflaton is slow-rolling [44]. In the following we calculate the first and the second terms separately.

D.1 Dynamically generated non-Gaussianities

We begin with calculating the second term,

$$\langle \mathcal{R}_c(\mathbf{k}_1)\mathcal{R}_c(\mathbf{k}_2)\mathcal{R}_c(\mathbf{k}_3)\rangle_{(2)}$$

$$\equiv \frac{1}{2}(N_{,\chi^2})^2 N_{,\chi^2,\chi^2}\left\langle [(\Delta\chi^2)^2]_{\mathbf{k}_1}(\delta\chi^2)_{\mathbf{k}_2}(\delta\chi^2)_{\mathbf{k}_3}+\text{c.p.}\right\rangle.$$
(D.3)

We have intentionally avoided calling this the dynamically generated bispectrum because, as we shall see below, it also includes contributions from the intrinsic bispectrum of $\delta\chi^2$.

On the other hand, noting that

$$\left[(\Delta\chi^2)^2\right]_{\mathbf{k}} = (\delta\chi^4)_{\mathbf{k}} - 2\langle\delta\chi^2\rangle(\delta\chi^2)_{\mathbf{k}},$$
(D.4)

we obtain

$$\left\langle \left[(\Delta\chi^2)^2\right]_{\mathbf{k}_1}(\delta\chi^2)_{\mathbf{k}_2}(\delta\chi^2)_{\mathbf{k}_3}+\text{c.p.}\right\rangle$$

$$= -2\times 3\times\langle\delta\chi^2\rangle\left\langle (\delta\chi^2)_{\mathbf{k}_1}(\delta\chi^2)_{\mathbf{k}_2}(\delta\chi^2)_{\mathbf{k}_3}\right\rangle$$

$$+ \left[\langle(\delta\chi^4)_{\mathbf{k}_1}(\delta\chi^2)_{\mathbf{k}_2}(\delta\chi^2)_{\mathbf{k}_3}\rangle+\text{c.p.}\right]$$
(D.5)

$$= -6\langle\delta\chi^2\rangle B_{\delta\chi^2}(k_1,k_2,k_3)(2\pi)^3\delta^3(\mathbf{k}_1+\mathbf{k}_2+\mathbf{k}_3)$$

$$+ \left[\langle(\delta\chi^4)_{\mathbf{k}_1}(\delta\chi^2)_{\mathbf{k}_2}(\delta\chi^2)_{\mathbf{k}_3}\rangle+\text{c.p.}\right].$$

The first term on the right-hand side above, containing $B_{\delta\chi^2}$, has the same form as the first term in Eq. (D.1) which we calculate later. We first concentrate on the other terms in the square brackets.

Note that if $\delta\chi^2$ were Gaussian, one could write these terms in terms of the product of two-point correlation functions. However, because $\delta\chi^2$ are non-Gaussian in nature, there is also a contribution proportional to the three-point correlation function of $\delta\chi^2$, as we shall see below.

Expanding $\delta\chi^4$ we obtain

$$\left\langle (\delta\chi^4)_{\mathbf{k}_1}(\delta\chi^2)_{\mathbf{k}_2}(\delta\chi^2)_{\mathbf{k}_3}\right\rangle = \int \widetilde{d^3q}\,\left\langle (\delta\chi^2)_{\mathbf{k}_1-\mathbf{q}}(\delta\chi^2)_{\mathbf{q}}(\delta\chi^2)_{\mathbf{k}_2}(\delta\chi^2)_{\mathbf{k}_3}\right\rangle,$$

(D.6)

in which the notation $\widetilde{d^3q} = d^3q/(2\pi)^3$ has been introduced for simplicity. To calculate the right-hand side of this equation we need to classify possible contractions. Since we are not interested in tadpole-type graphs but only in irreducible graphs, not all possible contractions are allowed. First, contractions between the terms only within $(\delta\chi^2)_p$ are discarded. Second, contractions should not be closed just on the terms within $(\delta\chi^4)_{\mathbf{k}_1}$, corresponding to the first two terms on the right-hand side of the above equation. Therefore, we find that there are two different types of allowed contractions. The first type is the one in which there is one contraction between a pair of terms from $(\delta\chi^4)_{\mathbf{k}_1}$. The second type is the one in which there is no contraction between any pair of terms from $(\delta\chi^4)_{\mathbf{k}_1}$ themselves.

Let us start with the first type and count the number of allowed contractions. We have 4 options to choose a pair from $(\delta\chi^4)_{\mathbf{k}_1}$. Then one of the remaining two $\delta\chi$ has 4 choices to contract with any $\delta\chi$ in $(\delta\chi^2)_{\mathbf{k}_2}$ and $(\delta\chi^2)_{\mathbf{k}_3}$, while the last $\delta\chi$ in $(\delta\chi^4)_{\mathbf{k}_1}$ has 2 choices to contract with the remaining terms in $(\delta\chi^2)_{\mathbf{k}_2}$ and $(\delta\chi^2)_{\mathbf{k}_3}$. In conclusion there are $4 \times 4 \times 2 = 32$ possible contractions in total in the first type, which are all equal. Therefore we calculate just one of them:

$$\int \widetilde{d^3q} \prod_i \widetilde{d^3p_i}$$

$$\left\langle \left(\delta\chi_{\mathbf{k}_1-\mathbf{p}_1-\mathbf{q}}\delta\chi_{\mathbf{p}_1}\right)\left(\delta\chi_{\mathbf{q}-\mathbf{p}_2}\delta\chi_{\mathbf{p}_2}\right)\left(\delta\chi_{\mathbf{k}_2-\mathbf{p}_3}\delta\chi_{\mathbf{p}_3}\right)\left(\delta\chi_{\mathbf{k}_3-\mathbf{p}_4}\delta\chi_{\mathbf{p}_4}\right)\right\rangle$$

$$= \int \widetilde{d^3q} \prod_i d^3p_i |\delta\chi_{|\mathbf{q}-\mathbf{p}_2|}|^2 |\delta\chi_{p_1}|^2 |\delta\chi_{p_2}|^2 |\delta\chi_{p_3}|^2 \times \delta\text{-factor},$$

$$(D.7)$$

in which

$\delta-$factor

$$= \delta^3\left(\mathbf{k}_1 - \mathbf{p}_1 - \mathbf{p}_2\right)\delta^3\left(\mathbf{k}_2 + \mathbf{p}_1 - \mathbf{p}_3\right)\delta^3\left(\mathbf{k}_3 + \mathbf{p}_2 - \mathbf{p}_4\right)\delta^3\left(\mathbf{p}_3 + \mathbf{p}_4\right).$$

$$(D.8)$$

Performing the integrals first over $\mathbf{p}_2$, $\mathbf{p}_3$ and $\mathbf{p}_4$, and then over $\mathbf{q}$, the above expression simplifies to

$$\int \widetilde{d^3q} \int d^3p_1 |\delta\chi_{|\mathbf{q}+\mathbf{p}_1-\mathbf{k}_1|}|^2 |\delta\chi_{p_1}|^2 |\delta\chi_{|\mathbf{k}_2+\mathbf{p}_1|}|^2 |\delta\chi_{|\mathbf{k}_1-\mathbf{p}_1|}|^2 \delta^3\left(\mathbf{k}_1 + \mathbf{k}_2 + \mathbf{k}_3\right)$$

$$= \langle\delta\chi^2\rangle \int d^3p_1 |\delta\chi_{p_1}|^2 |\delta\chi_{|\mathbf{k}_2+\mathbf{p}_1|}|^2 |\delta\chi_{|\mathbf{k}_1-\mathbf{p}_1|}|^2 \delta^3\left(\mathbf{k}_1 + \mathbf{k}_2 + \mathbf{k}_3\right)$$

$$= \frac{1}{8}\langle\delta\chi^2\rangle \left\langle (\delta\chi^2)_{\mathbf{k}_1}(\delta\chi^2)_{\mathbf{k}_2}(\delta\chi^2)_{\mathbf{k}_3}\right\rangle,$$

$$(D.9)$$

in which we have used the following identity [87]

$$\left\langle (\delta\chi^2)_{\mathbf{k}_1}(\delta\chi^2)_{\mathbf{k}_2}(\delta\chi^2)_{\mathbf{k}_3}\right\rangle$$

$$= 8\int d^3p_1 |\delta\chi_{p_1}|^2 |\delta\chi_{|\mathbf{k}_2+\mathbf{p}_1|}|^2 |\delta\chi_{|\mathbf{k}_1-\mathbf{p}_1|}|^2 \delta^3\left(\mathbf{k}_1 + \mathbf{k}_2 + \mathbf{k}_3\right).$$

Putting all together, we have

$$\int \widetilde{d^3q} \prod_i \widetilde{d^3p_i}$$

$$\left\langle \left(\delta\chi_{\mathbf{k}_1-\mathbf{p}_1-\mathbf{q}}\delta\chi_{\mathbf{p}_1}\right)\left(\delta\chi_{\mathbf{q}-\mathbf{p}_2}\delta\chi_{\mathbf{p}_2}\right)\left(\delta\chi_{\mathbf{k}_2-\mathbf{p}_3}\delta\chi_{\mathbf{p}_3}\right)\left(\delta\chi_{\mathbf{k}_3-\mathbf{p}_4}\delta\chi_{\mathbf{p}_4}\right)\right\rangle$$

$$= \frac{1}{8}\langle\delta\chi^2\rangle B_{\delta\chi^2}\left(k_1,k_2,k_3\right)(2\pi)^3\delta^3\left(\mathbf{k}_1+\mathbf{k}_2+\mathbf{k}_3\right),$$

$$(D.10)$$

in which the bispectrum $B_{\delta\chi^2}(k_1,k_2,k_3)$ is defined in Eq. (D.2). Because there are 32×3 of the same terms, the contribution from this type is given by

$$\left[\left\langle(\delta\chi^4)_{\mathbf{k}_1}(\delta\chi^2)_{\mathbf{k}_2}(\delta\chi^2)_{\mathbf{k}_3}\right\rangle + \text{c.p.}\right]_{1\text{st}}$$

$$= \frac{32 \times 3}{8}\langle\delta\chi^2\rangle B_{\delta\chi^2}\left(k_1,k_2,k_3\right)(2\pi)^3\delta^3\left(\mathbf{k}_1+\mathbf{k}_2+\mathbf{k}_3\right)$$

$$= 12\langle\delta\chi^2\rangle B_{\delta\chi^2}\left(k_1,k_2,k_3\right)(2\pi)^3\delta^3\left(\mathbf{k}_1+\mathbf{k}_2+\mathbf{k}_3\right)$$

$$(D.11)$$

Now let us look at the second type in which every $\delta\chi$ in $(\delta\chi^4)_{\mathbf{k}_1}$ is contracted with one $\delta\chi$ in either $(\delta\chi^2)_{\mathbf{k}_2}$ or $(\delta\chi^2)_{\mathbf{k}_3}$, so there are $4 \times 3 \times 2 = 24$ possible contractions. Again all these contractions yield the same result so we calculate only one of them:

$$\int \widetilde{d^3q} \prod_i \widetilde{d^3p_i}$$

$$\left\langle \left(\delta\chi_{\mathbf{k}_1-\mathbf{p}_1-\mathbf{q}}\delta\chi_{\mathbf{p}_1}\right)\left(\delta\chi_{\mathbf{q}-\mathbf{p}_2}\delta\chi_{\mathbf{p}_2}\right)\left(\delta\chi_{\mathbf{k}_2-\mathbf{p}_3}\delta\chi_{\mathbf{p}_3}\right)\left(\delta\chi_{\mathbf{k}_3-\mathbf{p}_4}\delta\chi_{\mathbf{p}_4}\right)\right\rangle$$

$$= \int \widetilde{d^3q} \prod_i d^3p_i |\delta\chi_{|\mathbf{k}_2-\mathbf{p}_3|}|^2|\delta\chi_{p_1}|^2|\delta\chi_{p_2}|^2|\delta\chi_{p_3}|^2 \times \delta\text{-factor},$$

$$(D.12)$$

in which

$$\delta\text{-factor} = \delta^3\left(\mathbf{k}_1+\mathbf{k}_2-\mathbf{p}_1-\mathbf{p}_3-\mathbf{q}\right)$$

$$\times \delta^3\left(\mathbf{k}_3+\mathbf{p}_1-\mathbf{p}_4\right)\delta^3\left(\mathbf{q}-\mathbf{p}_2+\mathbf{p}_3\right)\delta^3\left(\mathbf{p}_2+\mathbf{p}_4\right).$$

$$(D.13)$$

Performing the integrals first over $\mathbf{p}_2$, $\mathbf{p}_3$ and $\mathbf{p}_4$, and then over $\mathbf{q}$ yields

$$\int \widetilde{d^3 p_1} |\delta\chi_{p_1}|^2 |\delta\chi_{|\mathbf{k}_3+\mathbf{p}_1|}|^2$$

$$\times \int d^3 q |\delta\chi_{|\mathbf{q}+\mathbf{p}_1-\mathbf{k}_1|}|^2 |\delta\chi_{|\mathbf{q}+\mathbf{p}_1+\mathbf{k}_3|}|^2 \delta^3 (\mathbf{k}_1 + \mathbf{k}_2 + \mathbf{k}_3) \qquad \text{(D.14)}$$

$$= \frac{1}{4} P_{\delta\chi^2} (k_2) P_{\delta\chi^2} (k_3) (2\pi)^3 \delta^3 (\mathbf{k}_1 + \mathbf{k}_2 + \mathbf{k}_3) ,$$

where the definition of $P_{\delta\chi^2}$ given in Eq. (6.64) has been used. Since we have 24 of them plus the cyclic permutations, the total contribution from this type is given by

$$\left[\left\langle (\delta\chi^4)_{\mathbf{k}_1} (\delta\chi^2)_{\mathbf{k}_2} (\delta\chi^2)_{\mathbf{k}_3} \right\rangle + \text{c.p.} \right]_{2\text{nd}}$$

$$= 24 \times \frac{1}{4} \left[P_{\delta\chi^2} (k_2) P_{\delta\chi^2} (k_3) + \text{c.p.} \right] (2\pi)^3 \delta^3 (\mathbf{k}_1 + \mathbf{k}_2 + \mathbf{k}_3)$$

$$= 6 \left[P_{\delta\chi^2} (k_1) P_{\delta\chi^2} (k_2) + \text{c.p.} \right] (2\pi)^3 \delta^3 (\mathbf{k}_1 + \mathbf{k}_2 + \mathbf{k}_3) .$$

$$\text{(D.15)}$$

Adding up the results from Eqs. (D.11) and (D.15) together with the first term in (D.5) yields

$$\left\langle \left[(\Delta\chi^2)^2 \right]_{\mathbf{k}_1} (\delta\chi^2)_{\mathbf{k}_2} (\delta\chi^2)_{\mathbf{k}_3} + \text{c.p.} \right\rangle$$

$$= 6 \left(\langle \delta\chi^2 \rangle B_{\delta\chi^2} (k_1, k_2, k_3) \right.$$

$$\left. + \left[P_{\delta\chi^2} (k_1) P_{\delta\chi^2} (k_2) + \text{c.p.} \right] \right) (2\pi)^3 \delta^3 (\mathbf{k}_1 + \mathbf{k}_2 + \mathbf{k}_3)$$

$$\text{(D.16)}$$

Plugging the above result in Eq. (D.3), the three-point correlation function of $\mathcal{R}_c$ can be expressed in terms of the power spectrum and bispectrum of $\delta\chi^2$ as follows

$$\langle \mathcal{R}_c(\mathbf{k}_1)\mathcal{R}_c(\mathbf{k}_2)\mathcal{R}_c(\mathbf{k}_3)\rangle_{(2)}$$
$$= (2\pi)^3 (\mathbf{k}_1 + \mathbf{k}_2 + \mathbf{k}_3) \left[3 \left(N, \chi^2 \right)^2 N_{,\chi^2\chi^2} \langle \delta\chi^2 \rangle B_{\delta\chi^2} (k_1, k_2, k_3) \right.$$
$$\left. + 3 \left(N, \chi^2 \right)^2 N_{,\chi^2\chi^2} \left(P_{\delta\chi^2} (k_1) \, P_{\delta\chi^2} (k_2) + \text{c.p.} \right) \right].$$

$$(D.17)$$

As it is clear from the above result, the second term has the form of an ordinary non-linear interaction term with the vertex proportional to $N_{,\chi^2\chi^2}$ which can be easily evaluated. However, the first term is due to the intrinsic non-Gaussianity of the $\delta\chi^2$ field which has to be computed.

The above result can be further simplified by using Eq. (E.4) and noting that $P_{\mathcal{R}_c}^{wf} = N_{,\chi^2}^2 P_{\delta\chi^2}$, yielding

$$\langle \mathcal{R}_c(\mathbf{k}_1)\mathcal{R}_c(\mathbf{k}_2)\mathcal{R}_c(\mathbf{k}_3)\rangle_{(2)}$$
$$= \langle \mathcal{R}_c(\mathbf{k}_1)\mathcal{R}_c(\mathbf{k}_2)\mathcal{R}_c(\mathbf{k}_3)\rangle_{dyn} \qquad (D.18)$$
$$- 3(N\chi^2)^3 B_{\delta\chi^2} (k_1, k_2, k_3) \times (2\pi)^3 \delta^3 (\mathbf{k}_1 + \mathbf{k}_2 + \mathbf{k}_3) ,$$

in which the dynamically generated bispectrum of the curvature perturbation is defined via

$$\langle \mathcal{R}_c(\mathbf{k}_1)\mathcal{R}_c(\mathbf{k}_2)\mathcal{R}_c(\mathbf{k}_3)\rangle_{dyn}$$
$$= B_{\mathcal{R}_c}^{dyn} (\mathbf{k}_1, \mathbf{k}_2, \mathbf{k}_3) (2\pi)^3 \delta^3 (\mathbf{k}_1 + \mathbf{k}_2 + \mathbf{k}_3)$$
$$\equiv 3 \frac{N_{,\chi^2\chi^2}}{(N_{,\chi^2})^2} \left(P_{\mathcal{R}_c}^{wf} (k_1) P_{\mathcal{R}_c}^{wf} (k_2) + \text{c.p.} \right) (2\pi)^3 \delta^3 (\mathbf{k}_1 + \mathbf{k}_2 + \mathbf{k}_3) .$$

$$(D.19)$$

D.2 Bispectrum from intrinsic non-Gaussianity

Now we evaluate the bispectrum generated from the intrinsic non-Gaussian nature of $\delta\chi^2$. From Eqs. (D.1) and (D.18), we obtain

$$\langle \mathcal{R}_c(\mathbf{k_1})\mathcal{R}_c(\mathbf{k_2})\mathcal{R}_c(\mathbf{k_3})\rangle_{int}$$

$$= B_{\mathcal{R}_c}^{int}(\mathbf{k_1},\mathbf{k_2},\mathbf{k_3})(2\pi)^3\delta(\mathbf{k_1}+\mathbf{k_2}+\mathbf{k_3})$$

$$\equiv -2(N_{,\chi^2})^3 B_{\delta\chi^2}(\mathbf{k_1},\mathbf{k_2},\mathbf{k_3}) \times (2\pi)^3\delta^3(\mathbf{k_1}+\mathbf{k_2}+\mathbf{k_3}).$$

$$(D.20)$$

Following Ref. [87], we have the following expression for the bispectrum of $\delta\chi^2$

$$B_{\delta\chi^2}(k_1,k_2,k_3) = 8\int \widetilde{d^3q} |\delta\chi_\mathbf{q}|^2 |\delta\chi_{|\mathbf{k_1}-\mathbf{q}|}|^2 |\delta\chi_{|\mathbf{k_2}+\mathbf{q}|}|^2. \quad (D.21)$$

It is not easy to calculate the above integral in general, but it can be evaluated in the squeezed limit in which $k_3 \ll k_1 = k_2 \equiv k$. In this limit, the above integral simplifies to

$$B_{\delta\chi^2}^{\mathrm{sq}}(k) \equiv B_{\delta\chi^2}(k,k,0) \simeq 8\int \widetilde{d^3q}|\delta\chi_\mathbf{q}|^2|\delta\chi_{|\mathbf{k}-\mathbf{q}|}|^4. \quad (D.22)$$

In Appendix C we have calculated this integral and the result (given by Eq. (C.14)) is obtained to be

$$B_{\delta\chi^2}^{\mathrm{sq}}(k) \simeq \frac{\xi'^3}{k_c^3}\frac{H^4}{2\pi^2}P_{\delta\chi}(k) \simeq \left(\frac{\xi'}{\xi}\right)^3\frac{2H^2}{k_c^3}P_{\delta\chi^2}(k), \quad (D.23)$$

in which the second step follows from Eq. (C.10) or (6.67) and ξ and ξ' are two numerical constants of order unity.

Another limiting case of interest is the equilateral limit in which the amplitudes of all momenta are equal, $k_1 = k_2 = k_3$. Although we have no clue as to how to calculate the bispectrum in this limit, it is plausible that the amplitude of bispectrum is at most the same order as that in the squeezed limit, if not smaller, so we set

$$B_{\delta\chi^2}^{\mathrm{eq}}(k) = \xi^{\mathrm{eq}}\frac{2H^2}{k_c^3}P_{\delta\chi^2}(k), \quad (D.24)$$

in which ξ^{eq} is a dimensionless numerical parameter which can be of order unity.

D.3 Total f_{NL} parameter

As in previous chapters, we would like to calculate the non-Gaussianity parameter f_{NL} related to bispectrum via

$$\frac{6}{5} f_{NL}(k_1, k_2, k_3) = \frac{B_{\mathcal{R}_c}(k_1, k_2, k_3)}{\left[P_{\mathcal{R}_c}(k_1) P_{\mathcal{R}_c}(k_2) + \mathrm{c.p.} \right]}. \tag{D.25}$$

In our case it can be decomposed into three distinct contributions

$$f_{NL} = f_{NL}^{int} + f_{NL}^{dyn} + f_{NL}^{\phi}, \tag{D.26}$$

in which each term is defined via

$$\begin{aligned}
\frac{6}{5} f_{NL}^{int} &= \frac{B_{\mathcal{R}_c}^{int}(k_1, k_2, k_3)}{\left[P_{\mathcal{R}_c}(k_1) P_{\mathcal{R}_c}(k_2) + \mathrm{c.p.} \right]} \\
&= -2(N_{,\chi^2})^3 \frac{B_{\delta\chi^2}(k_1, k_2, k_3)}{\left[P_{\mathcal{R}_c}(k_1) P_{\mathcal{R}_c}(k_2) + \mathrm{c.p.} \right]},
\end{aligned} \tag{D.27}$$

$$\begin{aligned}
\frac{6}{5} f_{NL}^{dyn} &= \frac{B_{\mathcal{R}_c}^{dyn}(k_1, k_2, k_3)}{\left[P_{\mathcal{R}_c}(k_1) P_{\mathcal{R}_c}(k_2) + \mathrm{c.p.} \right]} \\
&= 3 \frac{N_{,\chi^2\chi^2}}{(N_{,\chi^2})^2} \frac{\left[P_{\mathcal{R}_c}^{wf}(k_1) P_{\mathcal{R}_c}^{wf}(k_2) + \mathrm{c.p.} \right]}{\left[P_{\mathcal{R}_c}(k_1) P_{\mathcal{R}_c}(k_2) + \mathrm{c.p.} \right]},
\end{aligned} \tag{D.28}$$

$$\begin{aligned}
\frac{6}{5} f_{NL}^{\phi} &= \frac{B_{\mathcal{R}_c}^{\phi}(k_1, k_2, k_3)}{\left[P_{\mathcal{R}_c}(k_1) P_{\mathcal{R}_c}(k_2) + \mathrm{c.p.} \right]} \\
&= \frac{1}{2} \frac{N_{,\phi\phi}}{(N_{,\phi})^2} \frac{\left[P_{\mathcal{R}_c}^{\phi}(k_1) P_{\mathcal{R}_c}^{\phi}(k_2) + \mathrm{c.p.} \right]}{\left[P_{\mathcal{R}_c}(k_1) P_{\mathcal{R}_c}(k_2) + \mathrm{c.p.} \right]}.
\end{aligned} \tag{D.29}$$

As noted before, the contribution of inflaton f_{NL}^{ϕ} is negligible, at most of the order of the slow-roll parameters, $f_{NL}^{\phi} = O(\epsilon, \eta)$ [44], so it can be safely ignored. Therefore, we concentrate on the contributions from $\delta\chi$ fluctuations.

First let us look at f_{NL}^{dyn}. As it is clear from its form, this is non-negligible only when the amplitude of $P_{\mathcal{R}_c}^{wf}(k)$ is comparable to or greater than that of $P_{\mathcal{R}_c}^{\phi}(k)$, and since the feature is highly localized, this happens only around the peak of the power spectrum at $k = k_{max}$. Then it is easy to check that f_{NL}^{dyn} is non-negligible only when k_1, k_2 and k_3 are all nearly equal to k_{max}. Therefore, setting $k_1 = k_2 = k_3 = k$ and in addition assuming that $P_{\mathcal{R}_c}^{wf}(k)$ dominates the power spectrum, we obtain

$$\frac{6}{5}f_{NL}^{dyn}(k = k_{max}) \simeq 3\frac{N_{,\chi^2\chi^2}}{\left(N_{,\chi^2}\right)^2} = -\frac{3\epsilon_\chi}{C\epsilon n_f^{1/2}}, \tag{D.30}$$

in which Eq. (E.4) has been used to eliminate $N_{,\chi^2}$ and $N_{,\chi^2\chi^2}$ from the intermediate expression. With $\epsilon_\chi \gg 1, C \ll 1$ and $n_f \sim 1$, we conclude that f_{NL}^{dyn} can be very large while centered at $k = k_{max}$. As an example, for the parameters used in our previous numerical example we obtain

$$\frac{6}{5}f_{NL}^{dyn}(k_{max}) \simeq 7 \times 10^4. \tag{D.31}$$

This is very large and may be excluded by the Planck data. However, this non-Gaussianity is highly localized so one cannot use the bounds from the Planck data trivially. One has to perform a new data analysis to put constraints on amplitude and location of this spiky non-Gaussianity.

Next we look at f_{NL}^{int}. As it is clear from the form of $B_{\delta\chi^2}$ given in Eq. (D.20), or by calculating it explicitly, it is finite in the squeezed limit, while the denominator of f_{NL}^{int} diverges because $P_{\mathcal{R}_c}^{\phi}$ scales like k^{-3}. As a result f_{NL}^{int} is completely negligible in the squeezed limit. However, the denominator is finite and scales like k^{-6} in the equilateral limit, so using Eq. (D.24), we can estimate

$$\frac{6}{5}f_{NL}^{int}(k) \simeq \xi^{eq}N_{,\chi^2}\frac{2H^2}{3k_c^3}\frac{P_{\mathcal{R}_c}^{wf}(k)}{P_{\mathcal{R}_c}^2(k)} = \xi^{eq}N_{,\chi^2}\frac{2H^2}{3}\frac{k^3}{k_c^3}\frac{P_{\mathcal{R}_c}^{wf}(k)}{P_{\mathcal{R}_c}^2(k)}. \tag{D.32}$$

Again, this contribution to non-Gaussianity is exponentially negligible except around the position of the peak of the waterfall field spectrum. Consequently, assuming $\mathcal{P}^{wf}_{\mathcal{R}_c} \simeq \mathcal{P}_{\mathcal{R}_c}$, and using Eqs. (6.68), (6.69) and (E.4), the above approximate expression yields

$$f^{int}_{NL}(k_{max}) \sim \xi^{\mathrm{eq}} N_{,\chi^2} \frac{H^2 k^3}{k_c^3} \frac{1}{\mathcal{P}_{\mathcal{R}_c}(k)} \sim \xi^{\mathrm{eq}} \epsilon_\chi^2 f^{dyn}_{NL}(k_{max}) . \qquad (\mathrm{D}.33)$$

Depending on the value of the unknown numerical parameter ξ^{eq}, the intrinsic non-Gaussianity can be larger than the dynamical non-Gaussianity.

In any case, we estimate that the total value of f_{NL} is at least as large as

$$f_{NL}(k_{max}) = f^{int}_{NL} + f^{dyn}_{NL} \sim \frac{\epsilon_\chi}{C\epsilon\sqrt{n_f}} . \qquad (\mathrm{D}.34)$$

The width of the induced feature in bispectrum is the same as that of the power spectrum given in Eq. (6.37), i.e., $\sigma(k) \sim 0.4 k_{max}$, so it is fairly localized. This justifies our interpretation of the spiky nature of non-Gaussianity.

APPENDIX E

$\delta\mathcal{N}$ up to $\Delta\chi^4$

In order to calculate $\delta\mathcal{N}$ up to order $\Delta\chi^4$, it is enough to extend Eq. (6.52) to the next order in $\delta\chi^2$, yielding

$$\frac{\Delta\chi^2(n)}{\langle\delta\chi^2(n)\rangle} - \frac{1}{2}\frac{\Delta\chi^4(n)}{\langle\delta\chi^2(n)\rangle^2} = \frac{\delta\chi^2_{min}(n_f)}{\chi^2_{min}(n_f)} + 2f'(n)\delta n - 2f'(nf)\delta n_f.$$

(E.1)

This gives additional corrections to $\delta\mathcal{N}$ in Eq. (6.56) as follows

$$\delta\mathcal{N} = \left(\frac{\Delta\chi^2(n)}{\langle\delta\chi^2(n)\rangle} - \frac{1}{2}\frac{\Delta\chi^4(n)}{\langle\delta\chi^2(n)\rangle^2}\right)\left[\frac{\partial n}{\partial n_f}\frac{1}{-2f'(n_f) + n_f^{-1}}\right]$$
$$+ \frac{\phi\delta\phi(n)}{2M_P^2}\left[1 + \frac{C}{2} + C\epsilon n + \frac{2f'(n)}{-2f'(n_f) + n_f^{-1}}\frac{\partial n}{\partial n_f}\right]$$

(E.2)

For a sharp waterfall transition, $n_f \lesssim 1$, the above expression simplifies to

$$\delta\mathcal{N} = -\frac{C\epsilon n_f}{f'(n_f)}\left(\frac{\Delta\chi^2(n)}{\langle\delta\chi^2(n)\rangle} - \frac{1}{2}\frac{\Delta\chi^4(n)}{\langle\delta\chi^2(n)\rangle^2}\right)$$
$$+ \left[1 + \frac{C}{2} + C\epsilon(n - 2n_f)\right]\frac{\phi\delta\phi}{2M_P^2}.$$

(E.3)

Using the above formula, one finds the following relation between $N_{,\chi^2}$ and $N_{,\chi^2\chi^2}$:

$$N_{,\chi^2} = -\langle\delta\chi^2(0)\rangle N_{,\chi^2\chi^2} = \frac{-C\epsilon n_f}{f'(n_f)}\frac{1}{\langle\delta\chi^2(0)\rangle} = -\frac{C\epsilon}{n_f^{1/2}\epsilon_\chi}\frac{1}{\langle\delta\chi^2(0)\rangle},$$

$$(E.4)$$

in which the second equality follows noting that $f'(n) = \epsilon_\chi n^{1/2}$.

Bibliography

[1] P. A. R. Ade et al. Planck 2013 results. XVI. Cosmological parameters. *Astron. Astrophys.*, 571:A16, 2014.

[2] P. A. R. Ade et al. Planck 2015 results. *XIII. Cosmological parameters*. 2015.

[3] G. Hinshaw et al. Nine-Year Wilkinson Microwave Anisotropy Probe (WMAP) Observations: Cosmological Parameter Results. *Astrophys. J. Suppl.*, 208:19, 2013.

[4] S. Perlmutter et al. Measurements of Omega and Lambda from 42 high redshift supernovae. *Astrophys. J.*, 517:565–586, 1999.

[5] Adam G. Riess et al. Observational evidence from supernovae for an accelerating universe and a cosmological constant. *Astron. J.*, 116:1009–1038, 1998.

[6] Daniel J. Eisenstein et al. Detection of the baryon acoustic peak in the large-scale correlation function of SDSS luminous red galaxies. *Astrophys. J.*, 633:560–574, 2005.

[7] Max Tegmark et al. Cosmological parameters from SDSS and WMAP. *Phys. Rev.*, D69:103501, 2004.

[8] K. Sato. First Order Phase Transition of a Vacuum and Expansion of the Universe. *Mon. Not. Roy. Astron. Soc.*, 195:467–479, 1981.

[9] Alan H. Guth. The Inflationary Universe: A Possible Solution to the Horizon and Flatness Problems. *Phys. Rev.*, D23:347–356, 1981.

[10] Alexei A. Starobinsky. A New Type of Isotropic Cosmological Models Without Singularity. *Phys. Lett.*, B91:99–102, 1980.

[11] Eiichiro Komatsu and David N. Spergel. Acoustic signatures in the primary microwave background bispectrum. *Phys. Rev.*, D63:063002, 2001.

[12] Viatcheslav F. Mukhanov. Gravitational Instability of the Universe Filled with a Scalar Field. *JETP Lett.*, 41:493–496, 1985. [Pisma Zh.Eksp.Teor.Fiz.41:402-405,1985].

[13] Misao Sasaki. Large Scale Quantum Fluctuations in the Inflationary Universe. *Prog. Theor. Phys.*, 76:1036, 1986.

[14] Christopher Gordon, David Wands, Bruce A. Bassett, and Roy Maartens. Adiabatic and entropy perturbations from inflation. *Phys. Rev.*, D63:023506, 2001.

[15] David Wands, Karim A. Malik, David H. Lyth, and Andrew R. Liddle. A New approach to the evolution of cosmological perturbations on large scales. *Phys. Rev.*, D62:043527, 2000.

[16] Misao Sasaki and Ewan D. Stewart. A General analytic formula for the spectral index of the density perturbations produced during inflation. *Prog. Theor. Phys.*, 95:71–78, 1996.

[17] Misao Sasaki and Takahiro Tanaka. Superhorizon scale dynamics of multiscalar inflation. *Prog. Theor. Phys.*, 99:763–782, 1998.

[18] David H. Lyth, Karim A. Malik, and Misao Sasaki. A general proof of the conservation of the curvature perturbation. *JCAP*, 0505:004, 2005.

[19] Alexei A. Starobinsky. Multicomponent de Sitter (Inflationary) Stages and the Generation of Perturbations. *JETP Lett.*, 42: 152–155, 1985. [Pisma Zh. Eksp. Teor. Fiz.42,124(1985)].

[20] D. S. Salopek and J. R. Bond. Nonlinear evolution of long wavelength metric fluctuations in inflationary models. *Phys. Rev.*, D42:3936–3962, 1990.

[21] Yasusada Nambu and Atsushi Taruya. Application of gradient expansion to inflationary universe. *Class. Quant. Grav.*, 13: 705–714, 1996.

[22] Hideo Kodama and Misao Sasaki. *Cosmological Perturbation Theory. Prog. Theor. Phys. Suppl.*, 78:1–166, 1984.

[23] Viatcheslav Mukhanov. *Physical Foundations of Cosmology.* Cambridge University Press, 2005.

[24] Steven Weinberg. *Cosmology.* Oxford University Press, 2008.

[25] David H. Lyth and Andrew R. Liddle. *The primordial density perturbation: Cosmology, inflation and the origin of structure.* 2009.

[26] Bruce A. Bassett, Shinji Tsujikawa, and David Wands. Inflation dynamics and reheating. *Rev. Mod. Phys.*, 78:537–589, 2006.

[27] Daniel Baumann. *Inflation. In Physics of the large and the small, TASI 09, proceedings of the Theoretical Advanced Study Institute in Elementary Particle Physics, Boulder, Colorado, USA, 1-26 June 2009*, pages 523–686, 2011.

[28] Yi Wang. Inflation, Cosmic Perturbations and Non-Gaussianities. *Commun. Theor. Phys.*, 62:109–166, 2014.

[29] William H. Kinney. TASI Lectures on Inflation. 2009.

[30] Atsushi Naruko, Yu-ichi Takamizu, and Misao Sasaki. *Beyond δN formalism. PTEP*, 2013:043E01, 2013.

[31] Kenji Tomita. Primordial irregularities in the early universe. *Progress of Theoretical Physics*, 48(5):1503–1516, 1972.

[32] Nathalie Deruelle and David Langlois. Long wavelength iteration of Einstein's equations near a space-time singularity. *Phys. Rev.*, D52:2007–2019, 1995.

[33] Masaru Shibata and Misao Sasaki. Black hole formation in the Friedmann universe: Formulation and computation in numerical relativity. *Phys. Rev.*, D60:084002, 1999.

[34] Naonori S. Sugiyama, Eiichiro Komatsu, and Toshifumi Futamase. δN formalism. *Phys. Rev.*, D87(2):023530, 2013.

[35] David H. Lyth and David Wands. Conserved cosmological perturbations. *Phys. Rev.*, D68:103515, 2003.

[36] Hardy Godfrey Harold. *A Course of Pure Mathematics*. Cambridge University Press, 1908.

[37] N. D. Birrell and P. C. W. Davies. *Quantum Fields in Curved Space*. Cambridge University Press, 2008.

[38] Viatcheslav Mukhanov and Sergei Winitzki. *Introduction to quantum effects in gravity*. Cambridge University Press, 2007.

[39] P. A. R. Ade et al. Planck 2015 results. XX. Constraints on inflation. 2015.

[40] P. A. R. Ade et al. Planck 2013 Results. XXIV. Constraints on primordial non-Gaussianity. *Astron. Astrophys.*, 571:A24, 2014.

[41] P. A. R. Ade et al. Planck 2015 results. XVII. Constraints on primordial non-Gaussianity. 2015.

[42] Eiichiro Komatsu. Hunting for Primordial Non-Gaussianity in the Cosmic Microwave Background. *Class. Quant. Grav.*, 27:124010, 2010.

[43] Xingang Chen. Primordial Non-Gaussianities from Inflation Models. *Adv. Astron.*, 2010:638979, 2010.

[44] Juan Martin Maldacena. Non-Gaussian features of primordial fluctuations in single field inflationary models. *JHEP*, 05:013, 2003.

[45] Daniel Babich, Paolo Creminelli, and Matias Zaldarriaga. The Shape of non-Gaussianities. *JCAP*, 0408:009, 2004.

[46] Leonardo Senatore, Kendrick M. Smith, and Matias Zaldarriaga. Non-Gaussianities in Single Field Inflation and their Optimal Limits from the WMAP 5-year Data. *JCAP*, 1001:028, 2010.

[47] Kazuya Koyama. Non-Gaussianity of quantum fields during inflation. *Class. Quant. Grav.*, 27:124001, 2010.

[48] Xingang Chen, Min-xin Huang, Shamit Kachru, and Gary Shiu. Observational signatures and non-Gaussianities of general single field inflation. *JCAP*, 0701:002, 2007.

[49] Guillem Domenech, Jinn-Ouk Gong, and Misao Sasaki. Consistency relation and inflaton field redefinition in the δN formalism. *Phys. Lett.*, B769:413–417, 2017.

[50] Andrei D. Linde and Viatcheslav F. Mukhanov. Nongaussian isocurvature perturbations from inflation. *Phys. Rev.*, D56: 535–539, 1997.

[51] David H. Lyth and David Wands. Generating the curvature perturbation without an inflaton. *Phys. Lett.*, B524:5–14, 2002.

[52] Kari Enqvist and Martin S. Sloth. Adiabatic CMB perturbations in pre - big bang string cosmology. *Nucl. Phys.*, B626:395–409, 2002.

[53] Takeo Moroi and Tomo Takahashi. Effects of cosmological moduli fields on cosmic microwave background. *Phys. Lett.*, B522: 215–221, 2001. [Erratum: Phys. Lett.B539,303(2002)].

[54] Misao Sasaki. Multi-brid inflation and non-Gaussianity. *Prog. Theor. Phys.*, 120:159–174, 2008.

[55] Atsushi Naruko and Misao Sasaki. Large non-Gaussianity from multi-brid inflation. *Prog. Theor. Phys.*, 121:193–210, 2009.

[56] David Polarski and Alexei A. Starobinsky. Isocurvature perturbations in multiple inflationary models. *Phys. Rev.*, D50.

[57] Viatcheslav F. Mukhanov and Paul J. Steinhardt. Density perturbations in multifield inflationary models. *Phys. Lett.*, B422:52–60,
1998.

[58] Gia Dvali, Andrei Gruzinov, and Matias Zaldarriaga. A new
mechanism for generating density perturbations from inflation.
Phys. Rev., D69:023505, 2004.

[59] Lev Kofman. Probing string theory with modulated cosmological
fluctuations. 2003.

[60] David H. Lyth. Generating the curvature perturbation at the end
of inflation. *JCAP*, 0511:006, 2005.

[61] Andrei D. Linde. Hybrid inflation. *Phys. Rev.*, D49:748–754,
1994.

[62] Edmund J. Copeland, Andrew R. Liddle, David H. Lyth, Ewan D.
Stewart, and David Wands. False vacuum inflation with Einstein
gravity. *Phys. Rev.*, D49:6410–6433, 1994.

[63] Mohammad Hossein Namjoo, Hassan Firouzjahi, and Misao
Sasaki. Violation of non-Gaussianity consistency relation in a single field inflationary model. *Europhys. Lett.*, 101:39001, 2013.

[64] Xingang Chen, Hassan Firouzjahi, Mohammad Hossein Namjoo,
and Misao Sasaki. A Single Field Inflation Model with Large
Local Non-Gaussianity. *Europhys. Lett.*, 102:59001, 2013.

[65] Xingang Chen, Hassan Firouzjahi, Eiichiro Komatsu, Mohammad
Hossein Namjoo, and Misao Sasaki. In-in and δN calculations of
the bispectrum from non-attractor single-field inflation. *JCAP*,
1312:039, 2013.

[66] Xingang Chen, Hassan Firouzjahi, Mohammad Hossein Namjoo,
and Misao Sasaki. Fluid Inflation. *JCAP*, 1309:012, 2013.

[67] Paolo Creminelli and Matias Zaldarriaga. Single field consistency
relation for the 3-point function. *JCAP*, 0410:006, 2004.

[68] Yi-Fu Cai, Xingang Chen, Mohammad Hossein Namjoo, Misao
Sasaki, Dong-Gang Wang, and Ziwei Wang. Revisiting non-
Gaussianity from non-attractor inflation models. 2017.

[69] William H. Kinney. Horizon crossing and inflation with large eta.
Phys. Rev., D72:023515, 2005.

[70] Jerome Martin, Hayato Motohashi, and Teruaki Suyama. Ultra Slow-Roll Inflation and the non-Gaussianity Consistency Relation. *Phys. Rev.*, D87(2):023514, 2013.

[71] C. Armendariz-Picon, T. Damour, and Viatcheslav F. Mukhanov. k - inflation. *Phys. Lett.*, B458:209–218, 1999.

[72] Jaume Garriga and Viatcheslav F. Mukhanov. Perturbations in k-inflation. *Phys. Lett.*, B458:219–225, 1999.

[73] Mohsen Alishahiha, Eva Silverstein, and David Tong. *DBI in the sky. Phys. Rev.*, D70:123505, 2004.

[74] David Seery and James E. Lidsey. Primordial non-Gaussianities in single field inflation. *JCAP*, 0506:003, 2005.

[75] Bernardo Finelli, Garrett Goon, Enrico Pajer, and Luca Santoni. Soft Theorems For Shift-Symmetric Cosmologies. 2017.

[76] Rafael Bravo, Sander Mooij, Gonzalo A. Palma, and BastiÃn Pradenas. A generalized non-Gaussian consistency relation for single field inflation. 2017.

[77] Sander Mooij and Gonzalo A. Palma. Consistently violating the non-Gaussian consistency relation. *JCAP*, 1511(11):025, 2015.

[78] Mohammad Akhshik, Hassan Firouzjahi, and Sadra Jazayeri. Effective Field Theory of non-Attractor Inflation. *JCAP*, 1507(07):048, 2015.

[79] P. A. R. Ade et al. Planck 2013 results. XXII. Constraints on inflation. *Astron. Astrophys.*, 571:A22, 2014.

[80] P. A. R. Ade et al. Planck 2015 results. XVI. Isotropy and statistics of the CMB. 2015.

[81] Jennifer A. Adams, Bevan Cresswell, and Richard Easther. Inflationary perturbations from a potential with a step. *Phys. Rev.*, D64:123514, 2001.

[82] Xingang Chen, Richard Easther, and Eugene A. Lim. Large Non-Gaussianities in Single Field Inflation. *JCAP*, 0706:023, 2007.

[83] Xingang Chen, Richard Easther, and Eugene A. Lim. Generation and Characterization of Large Non-Gaussianities in Single Field Inflation. *JCAP*, 0804:010, 2008.

[84] Ali Akbar Abolhasani, Hassan Firouzjahi, Shahram Khosravi, and Misao Sasaki. Local Features with Large Spiky non-Gaussianities during Inflation. *JCAP*, 1211:012, 2012.

[85] Ali Akbar Abolhasani and Hassan Firouzjahi. No Large Scale Curvature Perturbations during Waterfall of Hybrid Inflation. *Phys. Rev.*, D83:063513, 2011.

[86] Ali Akbar Abolhasani, Hassan Firouzjahi, and Misao Sasaki. Curvature perturbation and waterfall dynamics in hybrid inflation. *JCAP*, 1110:015, 2011.

[87] Jinn-Ouk Gong and Misao Sasaki. Waterfall field in hybrid inflation and curvature perturbation. *JCAP*, 1103:028, 2011.

[88] Alan H. Guth and S. Y. Pi. Fluctuations in the New Inflationary Universe. *Phys. Rev. Lett.*, 49:1110–1113, 1982.

[89] David H. Lyth. The hybrid inflation waterfall and the primordial curvature perturbation. *JCAP*, 1205:022, 2012.

Printed in the United States
By Bookmasters